INTRODUCTION TO FOOD SCIENCE *FOR KIDS!*
A Kitchen-Based Workbook

INTRODUCTION TO FOOD SCIENCE *FOR KIDS!*

A Kitchen-Based Workbook

Dale W. Cox

AN **EDIBLE KNOWLEDGE**® SERIES WORKBOOK

Copyright © 2020 by Dale W. Cox. All Rights Reserved

All rights reserved. No part of this publication may be reproduced, distributed, or transmitted in any form or by any means, including photocopying, recording, or other electronic or mechanical methods, without the prior written permission of the publisher, except in the case of brief quotations embodied in critical reviews and certain other noncommercial uses permitted by copyright law. For permission requests, write to the publisher, addressed "Attention: Permissions Coordinator," at the address below.

Paperback ISBN: 978-1-948515-09-2
Ebook ISBN: 978-1-948515-10-8

Library of Congress Cataloging-in-Publication Data is available.

Published by Beakers & Bricks, LLC

Cover design by Glen M. Edelstein
Interior design by Dale W. Cox and Glen M. Edelstein

Photographs and illustrations by Dale W. Cox, unless otherwise specified

Edible Knowledge® logo art by LeAnne Cox and Glen Edelstein
Edible Knowledge® is a registered trademark of Beakers & Bricks, LLC
Printed in the United States of America

Beakers & Bricks, LLC
PO Box 1014
Asheboro, North Carolina 27204
www.beakersandbricks.com

Disclaimer/caution: The experiments in this workbook involve high temperatures and sharp knives. Please take every precaution to avoid injury.

CONTENTS

INTRODUCTION   1

FOOD SCIENCE PRIMER: WHAT'S FOOD MADE OF?   4

MESS #1: SOFT AND FRESH TO STALE AND OLD: BREAD, PLUS EGG MEN ATTACK!   6
- Background   6
- Items Needed   6
- Procedures   7
- The Edible Knowledge   11

MESS #2: CORNSTARCH INTO PUDDING? REALLY!   14
- Background   14
- Items Needed (For Parts 1, 2, and 3)   14
- Procedures   14
  - Part 1   14
    - The Edible Knowledge, Part 1   16
  - Part 2   17
    - The Edible Knowledge, Part 2   18
  - Part 3   18
    - The Edible Knowledge, Part 3   19

MESS #3: FUN WITH COWS AND WHIPPING CREAM! — 22
- Background — 22
- Items Needed — 22
- Procedures — 23
  - Part 1 - Making Real Whipped Cream — 23
    - The Edible Knowledge, Part 1 — 24
  - Part 2 - Canned Whipping Cream — 25
    - The Edible Knowledge, Part 2 — 25
  - Part 3 - Butter — 26
    - The Edible Knowledge, Part 3 — 28

MESS #4: EGG-FOAM-TASTIC! — 30
- Background — 30
- Items Needed — 30
- Procedures — 30
- The Edible Knowledge — 33

MESS #5: GREEN EGGS THAT SPIN! — 36
- Background — 36
- Items Needed — 36
- Procedures — 36
- The Edible Knowledge — 37

MESS #6: HUGE MARSHMALLOWS! — 40
- Background — 40
- Items Needed — 40
- Procedures — 40
- The Edible Knowledge — 42

MESS #7: CAKE FOR DINNER! (OK, MAYBE NOT…) — 44
- Background — 44
- Items Needed (For Parts 1 and 2) — 44
- Procedures — 45
  - Part 1: Creamed, No Baking Powder — 45
  - Part 2: Creamed, With Baking Powder — 47
- The Edible Knowledge — 50

MESS #8: GLUEY, GOOEY GLUTEN… IN OUR FOOD? — 52
- Background — 52
- Items Needed (For Parts 1 and 2) — 53
- Procedures — 53
  - Part 1 - Very Little Kneading — 53
  - Part 2 - Lots of Kneading — 55
- The Edible Knowledge — 57

MESS #9: BANG PUFFING! — 60
- Background — 60
- Items Needed — 60
- Procedures — 60
- The Edible Knowledge — 62

MESS #10: FERMENTATION IS EVERYWHERE! — 66
- Background — 66
- Items Needed — 67
- Procedures — 67
- The Edible Knowledge — 68

MESS #11: EATING BROWN APPLES? YUCK! — 74
- Background — 74
- Items Needed — 74
- Procedures — 75
- The Edible Knowledge — 77

MESS #12: WATER IS ACTIVE — 80
- Background — 80
- Items Needed — 80
- Procedures — 81
- The Edible Knowledge — 83

MESS #13: HAVE A COW! — 86
- Background — 86
- Items Needed — 86
- Procedures — 86
- The Edible Knowledge — 89

MESS #14: CREAM THAT'S ICEY AND SMOOTH…AND LUMPY?! — 90

- Background — 90
- Items Needed — 90
- Procedures — 90
- The Edible Knowledge — 94

MESS #15: HEY! PINEAPPLE IS EATING MY JELL-O®?! — 96

- Background — 96
- Items Needed — 96
- Procedures — 97
- The Edible Knowledge — 99

THE END - OR JUST THE BEGINNING? — 101

PICTURE AND ILLUSTRATION ATTRIBUTIONS — 102

INTRODUCTION

Hello! I am a food science professional. What is that? Well, you've come to the right place to find out! For twenty-three years, I've worked for companies such as the Kellogg company (Frosted Flakes®!), General Mills, Inc.(Cinnamon Toast Crunch!™), and Post® (Fruity PEBBLES™!). Sounds like being a food scientist is a great job, right? Yes! Most people haven't ever heard about careers in food science, but they're a lot of fun. We get to apply scientific principles to interesting food products every day, and immediately see exciting reactions in our kitchen laboratories. Food science explains the science of cooking—what actually happens that causes your food to do what it does while it's being prepared.

To get this kind of a job, I went to college for a bachelor of science, and then I studied for two more years to obtain a master's degree in food science.

My company, Beakers & Bricks, LLC, started publishing the Edible Knowledge®

series of workbooks in 2018. This collection teaches food science, engineering, and physics principles. The workbooks are intended to be used as a part of a homeschool curriculum, or just for something fun to do if you're interested in science. One of my goals is to provide in-depth ideas and experiments that fully explain the principles of food science. This kind of depth

is lacking in much of the homeschool curriculum. Because scientific principles need to be explained deeply so the student grasps them firmly, a student usually needs to be closer to their teenage years. The series was initially designed for kids ages ten and up.

However, while exhibiting my books at conventions over the last two years, many attendees expressed great interest in a food science workbook for younger children. In the beginning I worried about how much depth would be reasonable and appropriate, but I decided that writing a book that provided early "scaffolding"—a basic foundation to be built upon later—would be useful and enjoyable. I spent quite a bit of time determining how best to proceed, with the result now in your hands...and I'm quite excited about it!

This workbook is designed to be used by young children, around six to ten years old, with help from their parents, older teens, or a supervisor of some sort. Some experiments can be done alone, depending on the student's age and maturity.

However, some experiments require using an oven or the stovetop, and sharp knives. The supervisor must read through the procedures and determine which experiments their student(s) can safely handle on their own, and which need oversight.

Learning about food science is rewarding and interesting! Even if the supervisor decides all the experiments need to be done as a team, I suspect they'll find the experience as rewarding and interesting as the student, and perhaps even more so.

Enjoy this serving of Edible Knowledge®!

Dale W. Cox
Food Scientist
February 2020

FOOD SCIENCE PRIMER

WHAT'S FOOD MADE OF?

This workbook is intended as a fun introduction for younger children, so you'll need to understand the five components that make up all foods.

WATER

You know what water is and what it looks like! Did you know it's an amazing thing, though? One side has a positive charge and the other has a negative charge, kind of like a magnet, which makes water great at dissolving things. Water is very important for food.

CARBOHYDRATES

You probably have heard the word "carbs," and that's short for "carbohydrates." The simplest and smallest carbohydrates are sugars, like normal table sugar and other words you might see on packaging labels: glucose, sucrose, and fructose.

Carbs have another role besides being sweet. When you connect a bunch of small sugars together into long chains, they lose their sweetness and become what we call "starch." Most grains—like wheat, oats, and barley, and also root vegetables, like potatoes—contain a lot of starch. The white powder we know as "flour" is mostly starch.

We get most of our energy from carbohydrates. Regardless of the source, though, for our body to use the fuel from carbs, they mostly have to be in the form of glucose, or sugar.

FOOD SCIENCE PRIMER: WHAT'S FOOD MADE OF? 5

PROTEIN

Proteins are the building blocks of everything in nature, including the food we eat. Proteins are found in muscles and enzymes. We'll discuss more about muscles and enzymes later, but for now, you should know proteins are very important for food scientists to learn about. Proteins are made of amino acids, which have a lot of nitrogen and other components.

FAT

Oils, butter, and shortening are all examples of fats. Fats are made of carbon molecules strung together in chains of different lengths, with hydrogen atoms all along the way. Some chains are straight, and others have kinks or bends in them. Some fats have three chains, others two, and others only one. The number, length, and character (straight or bent) of the chains are what determine what the fat is like. For instance, some fats are solid at room temperature, such as shortening; others are liquid, like vegetable or olive oil.

ASH

What's left over in your fireplace or charcoal grill once all the wood or charcoal is gone? Ash! In the same way, if you burn a piece of toast (or anything else) so completely that it won't burn anymore, there's a small amount of gray powdery stuff left called "ash." Regardless of what material is burned, ash mostly consists of minerals.

Ok, that's enough to get you going. Let's get started!

MESS 1: SOFT AND FRESH TO STALE AND OLD: BREAD, PLUS EGG MEN ATTACK!

Background

Has your bread ever sat around for a while and become stale? Wait…let's back up a minute. First, let's talk about what becoming stale ("staling") means. Say you buy a fresh loaf of bread from your local bakery, or maybe you've even been lucky enough to have one baked for you! If you squeeze the loaf of bread a little, it's super soft and springs back immediately when you let go.

Now, let it sit for a few days and do that again. It's not so soft anymore, is it? It probably tastes different, too. Donuts are also not nearly as good at the day's end then when they're fresh in the morning. What's going on here?

Let's find out!

Items Needed

- 1 fresh loaf of sliced organic white bread from the store. See the Important Note!
- 2 freezer-rated quart- or gallon-sized zip-closure–type plastic bags
- A ballpoint pen
- 1 cookie cutter with deep sides, preferably deeper than the bread, that fits within the bread at least 1/2 inch from all sides (a small gingerbread shaped one makes this experiment fun!)
- 1 plastic container that can hold about 5 cups of water. An empty non-dairy whipped topping container works great!
- 1/2 cup measuring cup
- Water (from the tap is fine)
- 1 dinner plate
- 1 ruler, at least 12 inches long
- Paper and a pen or pencil

MESS #1: SOFT AND FRESH TO STALE AND OLD: BREAD, PLUS EGG MEN ATTACK!

Important! It's important to get the right kind of bread, otherwise the experiment won't work. White bread gives the most dramatic results. (I'll explain why later.) Make sure the bread you purchase doesn't have ingredients called DATEM, calcium sulfate, or calcium propionate. Most organic breads don't have those.

The best is an Italian bread, if you can find one without those ingredients. A bread with no or very little whole pieces of wheat works best. Also, make sure it's fresh by performing the squeeze test described above. Find the softest loaf in the bunch, but don't squeeze the loaves too hard. You haven't purchased them yet!

Bread Bag Methodology: ("Methodology" is a series of steps or procedures to do something.) Whenever you open or close any of these marked bags, make sure they are closed firmly. Try to remove as much air from them as possible before closing. This goes for the zip-closure–type plastic bags, as well as the bag the bread came in.

Procedures

1. Mark one bag "Frozen" and the other "Room Temperature," and also mark the date and time with the pen, taking care not to puncture the bags.

2. Remove 3 slices from the bag of bread, excluding the heels (first and last pieces of the loaf) and the 2 slices closest to them at each end, and place them in the zip-closure–type plastic bag marked "Frozen." Remove as much air as you can from the bag and zip it closed. Place the bag in the freezer where it won't be disturbed or smashed.

3. Again excluding the heel and the two slices closest to either end, take another 3 slices, place them in the bag marked "Room Temperature," and set them on a counter in a place where they won't be disturbed.

4. Make sure your cookie cutter, plastic bowl, measuring cup, and water are ready to go. You might find it convenient to have a large bowl with water to dip from near your setup.

5. Take a third slice of bread from the bag (no heels or slices close to the heel!) and place it on a large flat plate. Leave the rest of the loaf in the original bag the bread came in, but don't eat it! You'll need some of it in 2 weeks.

What are the heels of the bread? These are the two ends of the bread. Some people call them "crusts," and others call them "butts"! They are somewhat thicker than the other slices and help hold the moisture in the rest of the loaf. Many people throw the heels away! But there are many ways you can use them if you just don't like them for a sandwich or toast.
https://www.thekitchn.com/here-are-5-reasons-to-never-throw-away-the-heel-of-bread-229890

8 INTRODUCTION TO FOOD SCIENCE *FOR KIDS!*

6. Carefully place the cookie cutter onto the bread so that it doesn't sink in. Now place the bowl, centered and level, on the cutter's top. **Don't press down!** Place the ruler next to your setup and note where the bowl's top meets the ruler. Write the measurement.

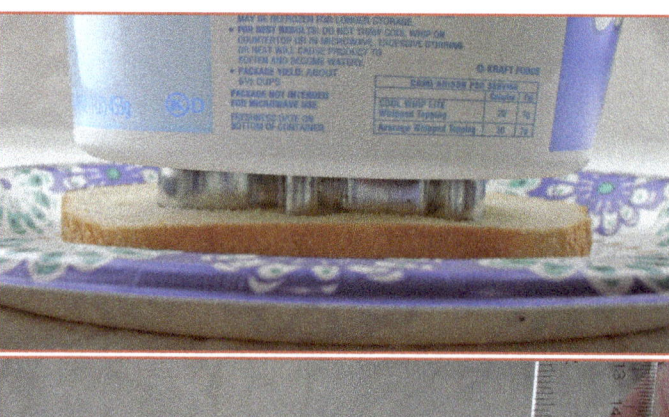

7. Add water (room temperature or cold, it doesn't really matter) 1/2 cup at a time, counting the number of cups you add to the bowl, waiting 30 seconds between each addition (the cutter will sink into the bread as time goes by, so you need to wait a little) and adjusting the bowl so it remains level. Measure again. Be careful. If the bowl isn't centered and level, you might have water spilled everywhere! Write down the number of cups it takes for the cutter to sink 3/8" into the bread.

8. Now take another fresh piece of bread from the bag (you know the drill by now). Use the same cookie cutter and cut out the slice's center. Eat this piece, observing what you notice with your senses: taste, smell, touch, hearing, and sight. (The bread is probably pretty quiet...) Write down your observations.

9. Remove the heels from the bag of bread and close the bag appropriately. You'll need the bag later. *(What do you do with the heels you took out? In my opinion, heels make the best toast...you should make some!)*

That's it for now! Set your cookie cutter and bowl aside. Put an X on your calendar for 10–12 days from today and wait until those days pass.

After 10 to 12 days

10. Take the bag out of the freezer and let it thaw overnight. The overnight thawing allows any water that has frozen on the bag's inside surface to melt and be reabsorbed by the bread.
11. Repeat Steps 7 and 8 above on the frozen and thawed bread, using the middle piece from the bag. Write down the number of cups of water it takes to "sink the cutter," and your observation.

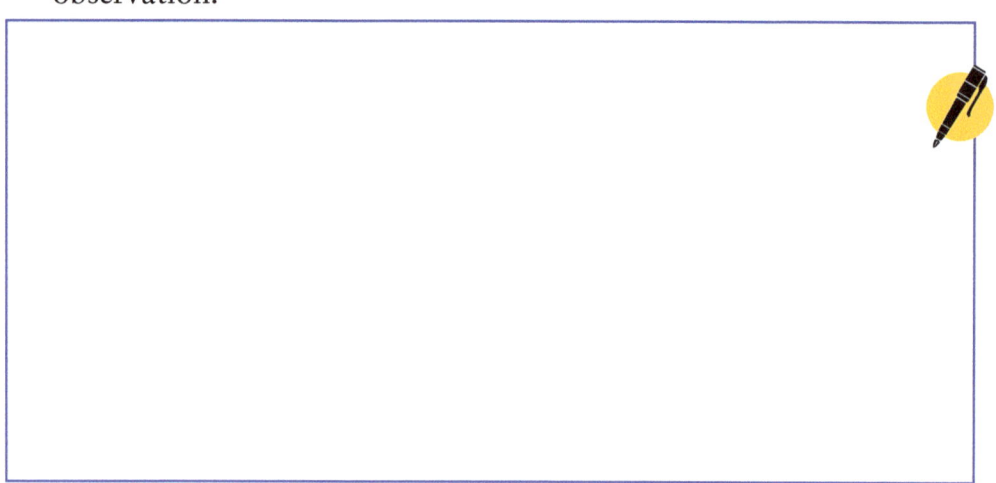

12. Repeat Steps 7 and 8 above on the room-temperature bread, using the loaf's center piece. Record the number of cups of water it takes to "sink the cutter," and your observations.

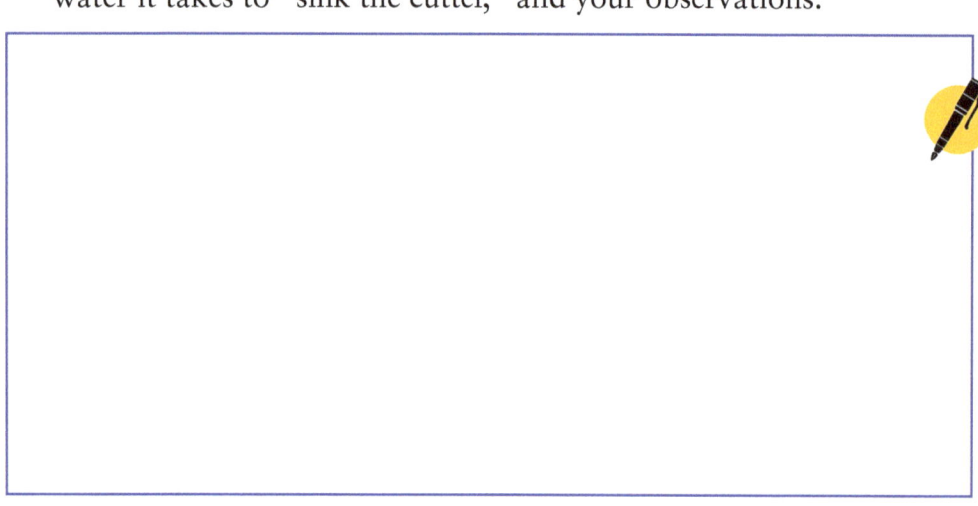

13. Repeat Steps 7 and 8 above on a piece of bread left in the original bag bread, using an outside slice. (Yes, this time use an outside slice.) Record the number of cups of water it takes to "sink the cutter," and your observations.

	Original Bag	Room Temperature	Frozen
Number of Cups			
Taste Test			

The Edible Knowledge

Baked goods are made of wheat flour, which is mostly starch. Remember that the white stuff in flour is mostly long chains of starchy carbohydrates. Baked goods, such as bread and donuts, become stale mostly for two reasons.

The first is due to moisture loss. When something dries out, it naturally gets harder and crispier. In our experiment, the zip-type bags are pretty good at preventing moisture loss, but the original bag the bread came in isn't as good.

The second reason the bread becomes stale is something called "starch retrogradation." When the ingredients of bread are mixed and the loaf of dough is put in the oven, the starch molecules warm up and relax their grip on each other, a part of "gelatinization," and become soft and stretchy, like a balloon. As soon as the perfectly baked loaf is removed from the oven, however, the soft and stretchy molecules want to hang onto each other again and become stiff. You can't stop starch retrogradation from happening, but you can slow it down. One way is by putting the loaf in the freezer.

One way you can think of retrogradation like toothpicks scattered randomly on a table, crossing one another in different directions. Over time (especially if you shake the table a bit) the toothpicks fall off each other and line up as much as they can. When starch is heated with water present, it's like dumping

> **One** way to make stale bread taste better is to toast it. You can try this on one of your room-temperature slices. As the bread is toasted, the starch molecules separate from one another again and the toast becomes very similar to fresh bread toast.

starch "toothpicks" on a table, ending up in that every-which-way pile. Retrogradation is like shaking the table, lining up the starch again. When they do line up, the bread gets tougher or firmer, and the flavor changes.

Toothpick starch retrogradation demonstration:
https://www.youtube.com/watch?v=4_dXFBom5EI

It took less effort, or weight, to squish the fresh bread with the bowl and cookie cutter. When you followed the directions, you may have noticed that you wrote down fewer half cups of water poured into the bowl sitting on top of the cookie cutter.

Freezing slows down the process of retrogradation, so when that bread is thawed, it's softer than the bread left out on the counter. The most firm is the bread left in the original bag. Those bags may look sturdy, but most of them are thin and aren't the best at preventing moisture loss.

If you come up with results that are different from what I describe, they might be due to the bread's plastic bag thickness (some are thin, some are thick), and also the zip-type bag's thickness. They can vary, and these measurements will have an effect on the experiment. Can you think of other reasons why your results might not be the same as mine?

Dough conditioners: DATEM, calcium sulfate, and calcium propionate all "condition" dough in different ways, such as making the dough softer, encouraging it to rise more readily, and increasing the time it remains soft after baking. Some commercial breads will stay soft for a very long time, which would make this experiment difficult.

The best bread to show retrogradation would be homemade bread, but there are too many variables (baking, slice thickness, etc.) to make that a good test.

MESS #1: SOFT AND FRESH TO STALE AND OLD: BREAD, PLUS EGG MEN ATTACK!

Ok, so what about the egg man? Take one of your pieces of bread with the cut-out hole in the middle. Spread some butter on what remains, both sides. Place in a frying pan set at a temperature to fry an egg. Now carefully add an egg in the cut-out space. After it sets up a bit, flip it. You'll see the shape outlined in egg white! These are fun and tasty to eat!

A video showing one method to make stale bread taste good again: https://www.youtube.com/watch?v=5X6JZLnP0kU

That's it! Now let's clean up the mess!

MESS 2: CORNSTARCH INTO PUDDING? REALLY!

Background

Have you ever made pudding? Lots of brands of ready-to-eat pudding are available to chow down on, but making it in your kitchen is fun. Let's try it!

Items Needed (For Parts 1, 2, and 3)

- 4 tablespoons cornstarch
- Water
- 2 containers with lids for use in the refrigerator
- 1/2 cup white sugar
- 1/8 teaspoon salt
- 2 3/4 cups milk (preferably whole milk)
- 2 tablespoons butter, room temperature
- 1 teaspoon salt
- 3 tablespoons unsweetened cocoa powder
- 1/2 teaspoon vanilla
- 1 small saucepan
- 1 large saucepan
- Masking or painter's tape for labels
- Marker or pen

PART 1

Procedures

1. Label one container "Cornstarch" and the other "Pudding."
2. In the small saucepan, add 3/4 cup water and 2 tablespoons cornstarch. Stir these together while cold.

MESS #2: CORNSTARCH INTO PUDDING? REALLY! 15

3. Begin heating slowly (on low heat) and note what happens to the ingredients in the saucepan. Increase the heat until the mixture boils. Let it boil for 1 minute, stirring constantly and gently. Remove from the heat.

4. While the mixture is still warm, fill the container marked "Cornstarch" with about 1/2 the blended goo. Take a picture so you'll be able to compare its appearance now to what it will look like a week from now, in Part 3. Set the "Cornstarch" container aside and let it cool to room temperature. Then affix the lid and place the container in the refrigerator.

> **Texture** refers to what something feels like. In this case, either what it feels like when stirring the mixture with a spoon, or in your mouth. Another term used by food scientists for what something feels like in your mouth is "mouthfeel." Isn't that a funny word?

5. Let the remaining mixture in the saucepan cool enough so you can taste it. Write down your observations. Pay attention to the *texture*.

16 INTRODUCTION TO FOOD SCIENCE *FOR KIDS!*

The Edible Knowledge, Part 1

So what's going on here? Cornstarch is used as a thickener in lots of things, like gravy and pudding.

How does it work to make liquids thicker?

Cornstarch is made up of granules of starch. When heated in plenty of water, the starch granules swell as the water soaks into them.

Think of these granules like balloons. You have fifty empty balloons and begin inflating them inside a small room. As they inflate, or swell, the room gets crowded! The balloons (the granules) start to bump into each other, making it harder for them to move around. Rubbing against each other is called "friction," and it's what makes the mixture thicken! This process is another example of starch gelatinization.

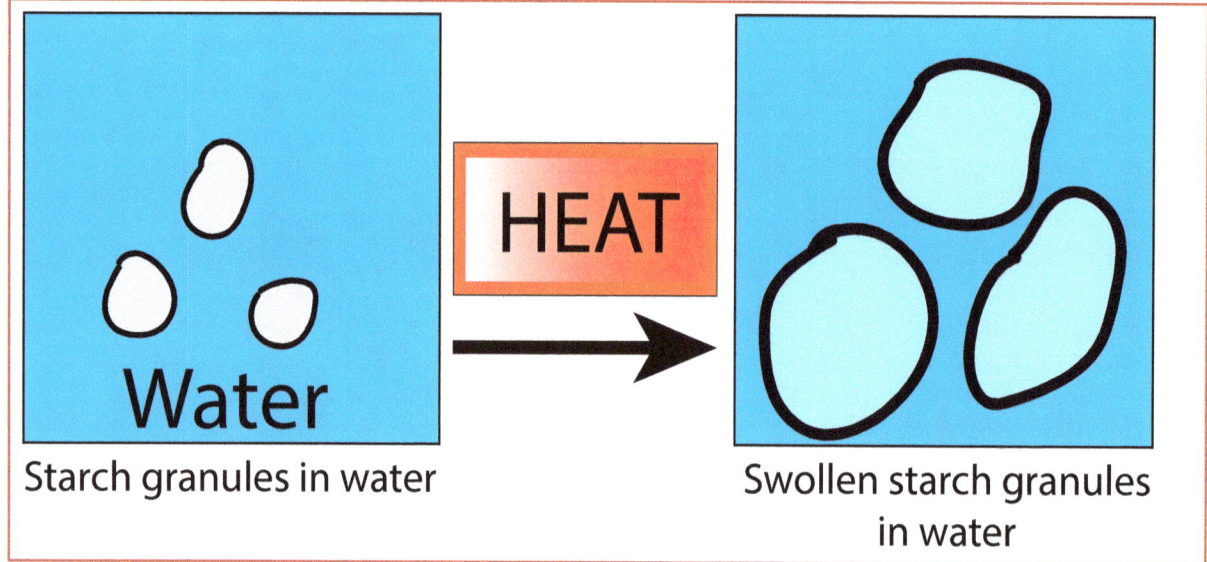

Starch granules in water   Swollen starch granules in water

Now for the next part. If you have a box of cook-on-stove pudding (NOT instant), look at the instructions. They usually say to "stir gently." Why?

Apart from preventing splashes of hot mixtures that can burn you, there's a food science reason: The starch balloons will pop if you stir too hard! When they pop, their insides sort of fall out and they get smaller again, resulting in a lot less friction. In our example, the result is that the pudding isn't as thick as it should be. You might have pudding soup!

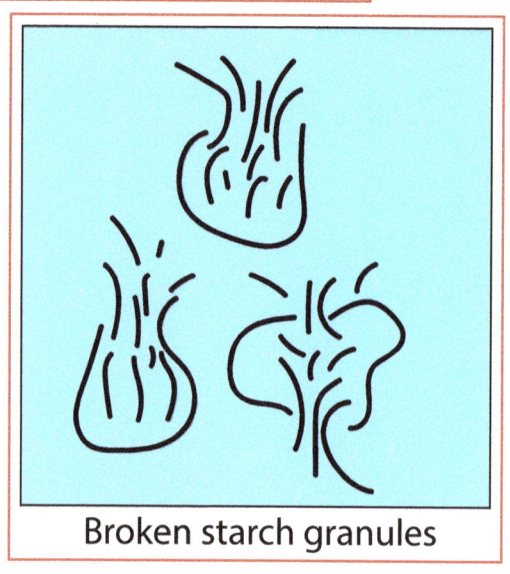

Broken starch granules

MESS #2: CORNSTARCH INTO PUDDING? REALLY! 17

PART 2

Now that you know how cornstarch pudding thickens, let's make something that tastes good!

1. In the large saucepan, combine the sugar, cocoa, 2 tablespoons cornstarch, and salt. Mix these together. Then stir in the milk.

2. Place the saucepan over medium heat, stirring constantly and gently until the mixture comes to a full boil for at least 1 whole minute.

3. Remove the saucepan from the heat and immediately stir in the butter and vanilla.

18 INTRODUCTION TO FOOD SCIENCE *FOR KIDS!*

4. Once the mixture has cooled a bit, add about the same amount to the refrigerator container labeled "Pudding" as you did in Part 1. Take a picture (with a camera, tablet, or smartphone) so you can compare it later in Part 3.
5. Let it cool to room temperature, put the lid on, and place the container in the fridge alongside the container holding the plain cornstarch.
6. Pour the rest into serving dishes of your choice.
7. Once it's cooled enough, taste the pudding and notice the texture. Write down your observations:

That's it for now! Put an X on your calendar 7–10 days from today and wait until those days pass.

The Edible Knowledge, Part 2

Did you notice the difference in texture? Apart from tasting better, sugar, cocoa, and milk all affect how starch granules gelatinize. These ingredients tend to slow down starch gelatinization. So it may have taken the mixture a little longer to thicken when it contained the extras.

PART 3

After 7 days

1. Remove the 2 containers from the refrigerator and remove the lids.
2. Observe both containers without disturbing them and record your observations, comparing them with what you see in the pictures you took last week.

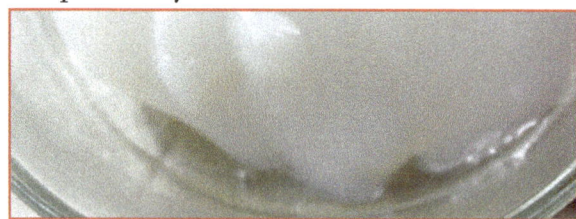

MESS #2: CORNSTARCH INTO PUDDING? REALLY! 19

3. Now taste each container, paying particular attention to the mouthfeel. Write down your observations.

The Edible Knowledge, Part 3

So, what did we see here? When you took the lids off, you likely noticed liquid on top of the mixtures, and maybe also on the bottom of the lids, but more on the plain cornstarch's top. When comparing the two containers to the pictures from the first week, you can see that there wasn't any of this liquid.

Where did it come from?

It's another example of retrogradation. The starch molecules in the pudding, over time, tend to reconnect with one another (it's the same thing that happened in the bread experiment!). This re-forms links that had been broken by cooking the ingredients. When this happens in something that contains a lot of liquid, these links pull together and shrink a little bit. This squeezes out the liquid. Another name that food scientists use for this squeezing and liquid loss is called syneresis.

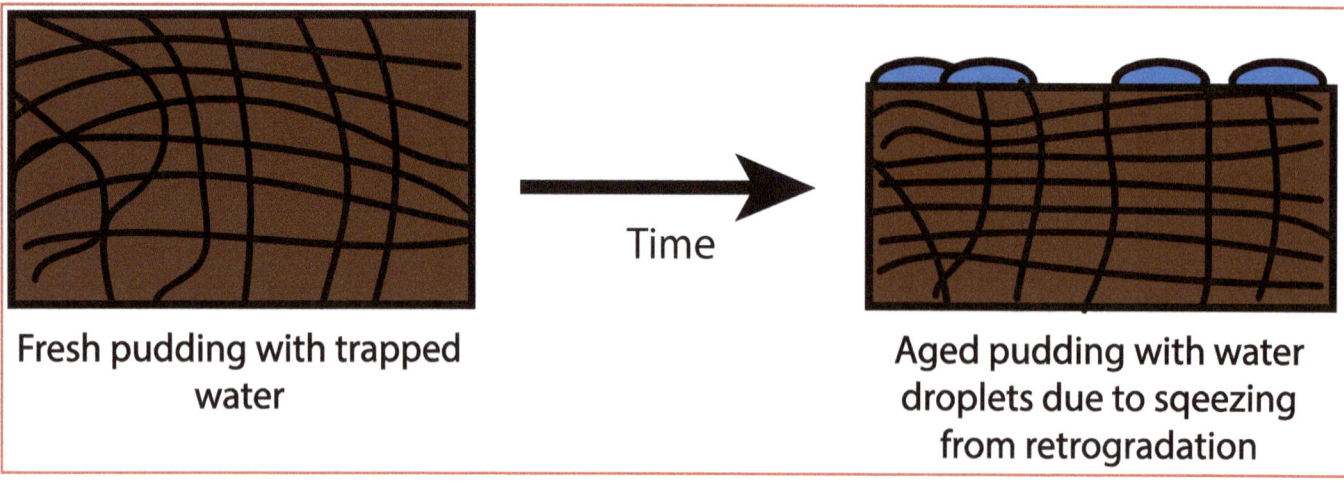

Fresh pudding with trapped water

Aged pudding with water droplets due to sqeezing from retrogradation

You likely saw more liquid on the plain cornstarch than on the chocolate pudding.

Why?

Just as the butter, milk, sugar, and other ingredients listed in Part 1, Step 2, delayed the cornstarch's gelatinization when making the pudding, they also slow down retrogradation and tend to hang on to the water when it's finished cooking. Both things prevent syneresis to some degree.

The chocolate pudding's finished texture is smoother than the plain cornstarch. The smoothness comes from less retrogradation, and also from the fat in the milk, cocoa, and butter. These additions improve the product's mouthfeel as well.

Food scientists can use their knowledge of ingredients to make pudding that's ready-to-eat and just right!

Video on starch gelatinization. Some vocabulary in this video is beyond the text of this workbook, but you'll get most of it. https://www.youtube.com/watch?v=oiGUyvMHqM4

That's it! Now let's clean up the mess!

Thanksgiving gravy[12]

Gravy is an example of a food thickened using starch.

MESS 3: FUN WITH COWS AND WHIPPING CREAM!

Background

We love this white stuff on pies! If you haven't tried it, you should.

But how does it work? How do you get light, fluffy stuff from cream, a heavy liquid?

Let's figure it out!

A Lazy Day[1]

Items Needed

- 1 cup heavy whipping cream
- 1 1/2 tablespoons powdered sugar (also called "confectioners' sugar")
- 1/4 teaspoon vanilla extract
- 1 can of original flavor whipping cream, such as ReddiWhip® (Keep it in the refrigerator for now.)
- 2 cold (refrigerated) mixing bowls—glass is best, and they must be very clean
- 1 cold (refrigerated) plate
- 1 set of candy molds, or something small (about 1 tablespoon) that can be used as a mold
- A handheld or countertop mixer with the beaters chilled (refrigerated)

> *Tip:* You only need to chill the beaters themselves—the part that comes into contact with the food. You can do this by placing them and the bowl and plate in the refrigerator for at least thirty minutes before starting this experiment.

PART 1 – MAKING REAL WHIPPED CREAM

Procedures

1. Pour the cream into one chilled bowl and beat on high speed until the mixture starts to thicken a little.
2. Remove about half the mixture and place it in the second bowl. Cover the second bowl with plastic wrap, or a lid, and put it in the refrigerator.

3. In the first bowl, sprinkle in the powdered sugar and vanilla and beat on high again until soft peaks form. "Soft peaks" means that when you pull the beaters out, small mounds stay in place and don't collapse, but the tip wilts.

4. Taste the whipping cream, but save some for comparison in Part 2. Record your observations:

The Edible Knowledge, Part 1

Proteins are amazing molecules, and there's a lot of protein (as well as fat!) in cream. Here we saw one protein property, which is the ability to create foams.

How does it work?

Proteins can be long molecules, but they are usually coiled up alone, or even with other proteins. However, they can be forced to do other things. In our case, we beat them with the mixers. Beating breaks up the associations—or connections—that proteins have between themselves and other proteins. They become uncoiled and stretched out.

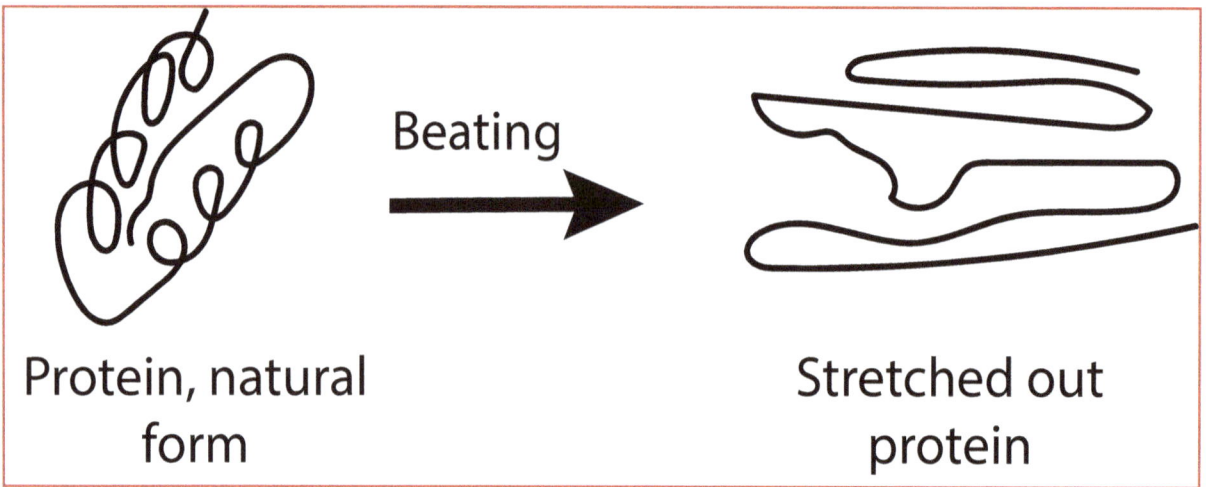

The stretched-out protein strands still interact with one another, but since they're stretched out, space exists between them, forming a loose matrix. In addition to stretching out the proteins, the process of beating with the mixer adds in air, which gets trapped in the stretched-out protein structure.

Pretty cool!

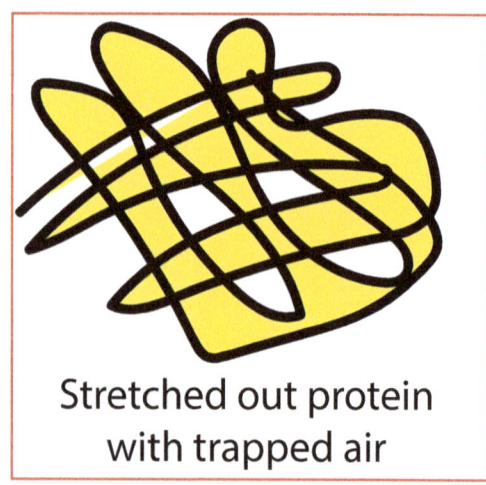

PART 2 – CANNED WHIPPING CREAM

1. Remove the can of whipping cream from the refrigerator. Shake the can, turn it upside down, and spray some on the plate.
2. Taste it, and compare the texture and flavor with the whipping cream you just made.
3. Look at the side panel on the can of whipping cream and read the ingredients. Record your observations and your thoughts about how the cream comes out already expanded:

The Edible Knowledge, Part 2

So, what's different here? When you looked at the side panel, you saw multiple ingredients listed, like cream, sugar, water, corn syrup (more sugar), and milk, along with flavor, mono- and diglycerides, and possibly something called carrageenan, with nitrous oxide used as a propellant. Your particular can might differ a little from this, but probably not much. Some of these ingredients cause the cream inside the can to change from liquid to foam instantly.

Let's look at them.
- Sugar and corn syrup provide sweetness and some flavor, along with any added flavor.
- Mono- and diglycerides are "emulsifiers." These help water and fat to mix, and stay that way. Have you ever tried to blend oil and water, like in salad dressing? They'll stay together briefly, but when left sitting still, they'll separate again. Emulsifiers help to prevent the separation.

26   INTRODUCTION TO FOOD SCIENCE *FOR KIDS!*

- Carrageenan is a carbohydrate that's long and complex, called a "hydrocolloid." These large molecules hold on to water and can be used to thicken foods.
- Nitrous oxide (NO2) is a gas that has been compressed (squished) into the can. That's why the can is said to be "under pressure."

OK, now we have the ingredients. The instructions say to shake the can, which mixes all the ingredients together, and the emulsifier will help them stay together. Hold the can so the nozzle is down, and spray it. The nozzle's shape and the can's tubing force the proteins in the cream to stretch out as they move along. The compressed nitrous oxide immediately expands as it exits with the now-stretched-out-and-interacting proteins, and some of the gas gets caught and trapped, creating the foam we see!

> **Carrageenan** is an extract of seaweed! It's common in dairy products (foods that contain milk) because of its ability to interact and create texture with dairy proteins. Look at the side of a box of ice cream; it's probably in there! So just think, when you're eating ice cream, you're eating seaweed. Cool!

PART 3 – BUTTER

What's the difference between whipped cream and butter? More whipping! Let's beat the crud out of this stuff.

1. Clean off your beaters if you haven't already. It's best to make sure they're cold again as well.
2. Remove the saved, slightly thickened whipped cream from the refrigerator.
3. Beat the mixture on high again, this time for a while. You'll see soft peaks forming, but you want to keep going for at least 3 minutes.
4. You will see the texture change, and liquid will start to come out. Keep beating it a little longer until it doesn't look like any more liquid is coming out. (This is called being "over-whipped," which most of the time is a bad thing!)

MESS #3: FUN WITH COWS AND WHIPPING CREAM! 27

5. Separate the liquid—which is buttermilk!—from the solids. Squeeze the solids together to force out most of the liquid.
6. If you want, you can mix a little salt into the solids, just a bit. Maybe 1/8 teaspoon.
7. Press the solids into the mold(s) and chill them in the refrigerator. If you only have one mold, you can form the butter in the mold and carefully remove it. Put it on a tray to cool in the fridge. That way you can reuse the mold as many times as needed.

You've just made butter!

Record your observations:

The Edible Knowledge, Part 3

Wow! You made butter. I'll bet you didn't know where buttermilk came from, but now you do! Buttermilk is great for making other foods, such as pancakes. It provides a rich flavor that many people like.

So how did the extra beating create butter? Back to proteins.

First we made a foam with our loose, stretched-out, and interacting proteins that were trapping gas (air), fat, and water. As we kept beating, these proteins become more and more abused, eventually losing their ability to form a loose mesh, at which point the gas and some water can escape. What's left is fat and a little protein, with some water mixed in. It's still an emulsion, but has flipped from oil in water (cream) to water in oil.

That's it! Now let's clean up the mess!

MESS #3: FUN WITH COWS AND WHIPPING CREAM!

Mooooooo...

Oh! I forgot the cow! What does a cow have to do with whipped cream and butter? Well, where do you think the cream comes from in the first place? (And don't say the grocery store!)

Cow's milk is mostly water, with some carbs and protein, but it also contains fat. When milk sits still overnight, the fat—also known as cream!—separates from the watery milk and floats to the top. Busy food scientists don't generally like to wait around for the cream to separate naturally, so they spin the milk around through a centrifuge to separate the cream.

I have a question for you: If you milk a brown cow, do you get chocolate milk?

That's it! Now let's clean up the mess!

An Australian guy making butter in his kitchen https://www.youtube.com/watch?v=jDq269e6w_c

See how milk gets from the barn to your table https://youtu.be/-sQS4PU3ZZE

MESS 4: EGG-FOAM-TASTIC!

Background

Eggs are sometimes called "the perfect protein" because they have a lot of nutrients. But from a food science perspective, eggs are also very cool.

Let's take a look at how we can make foams with them…and a couple other things.

Items Needed

- 2 whole eggs (in the shell, unbroken)
- 1/2 cup sugar
- 1 baking sheet for the oven
- Parchment paper
- 1 handheld or countertop mixer
- 1 medium-sized clean metal or glass bowl
- 1 medium-sized container with lid for the refrigerator

Procedures

1. Preheat the oven to 425°F.
2. Separate the eggs yolks from the egg whites. Make sure not to get any yolks in the whites! Keep the egg whites in the clean metal or glass bowl, and you can either throw out the egg yolks or feed them to your dogs—they'll love them! If you want to cook them, the internet has good yolk-only recipes. And if you don't know how to separate eggs, check out the videos.
3. Using the mixer, beat the egg whites until they form soft peaks.
4. Slowly add about 1/3 of the sugar and continue to

30

Some ways to separate egg whites and yolks. I always use the shell-to-shell method, but here's a look at a couple of novel ways:

https://www.youtube.com/watch?v=yAGX-54iR30

https://www.youtube.com/watch?v=X3tzp_Sws2g

mix. Slowly add in the second 1/3 and mix, and then the final 1/3.

5. Continue beating the mixture once all the sugar has been

added until firm peaks are achieved. When you pull the beaters straight up out of the beaten whites, firm peaks are taller and more sturdy than soft peaks.

6. Place a sheet of parchment paper on the baking sheet.

7. Scoop out approximately 4 mounds of egg whites, one at a time, and put the mounds on the baking sheet.

8. Place the baking sheet in the oven and watch carefully. Remove it when the surfaces of the egg whites begin to brown, or after about 10 minutes.
9. Remove the sheet from the oven and let the mounds cool. Egg whites that are baked after having been beaten are called "meringues." As they cool, you'll see them collapse a bit.

10. You need to save one in a container and place the container in the refrigerator. But the meringues are probably stuck to the parchment paper now! You might find it easiest to cut the parchment paper around the mound's bottom and put the whole thing—the meringue on the parchment paper—in the container. Record your observations:

That's it for now, but what do you do with the other three meringues? Well, you could…eat them! For flavoring, you could sprinkle some cinnamon sugar or brown sugar on top.

Then put an X on your calendar seven to ten days from today and wait until those days pass.

After 7 days

Remove the container from the refrigerator and observe the meringue for any changes. Do you see any beads of liquid on the surface? If so, taste some of it. Record your observations:

The Edible Knowledge

Because they're similar to the proteins in cream, egg white proteins are great for making foams. When the egg whites are beaten with the mixer, they lose their natural form and stretch out. They end up in a loose mesh form that can trap air, water, and sugar. Any change in shape of a protein is called *denaturation*.

Text Box: Denaturation refers to the loss of the normal function of a protein. The loss can be permanent or reversible. Usually when heated, protein denaturation is permanent.

However, we see something different when we bake the meringue! Instead of melting, its shape sets and becomes firm. What happens here? The egg white proteins continue to denature, but this time the change becomes permanent as they bind to neighboring proteins.

> ***Denaturation*** refers to the loss of the normal function of a protein. The loss can be permanent or reversible. Usually when heated, protein denaturation is permanent.

The meringues might collapse a little after removing them from the oven, as trapped air cools and contracts, but the egg whites hold most of their shape.

By contrast, you saw that the whipping cream foam lost its shape when it was left at room temperature even for a few minutes. However, beaten meringue eventually can spoil, so the unbaked meringue mixture shouldn't be left unrefrigerated for long periods of time.

When you checked the meringue left in the refrigerator, you likely saw little beads of liquid on the surface. These beads are an example of protein mesh shrinkage, or denaturing. That process squeezes out some of the liquid. It's a good example of syneresis.

Do you remember? When food squeezes itself so that it loses liquid, food scientists call the process *syneresis*.

In the case of meringues, the liquid being squeezed out is a combination of water and sugar, so the beads are sweet! Other ingredients can be added to meringues to stabilize them and prevent shrinkage, achieve different textures, and produce other changes.

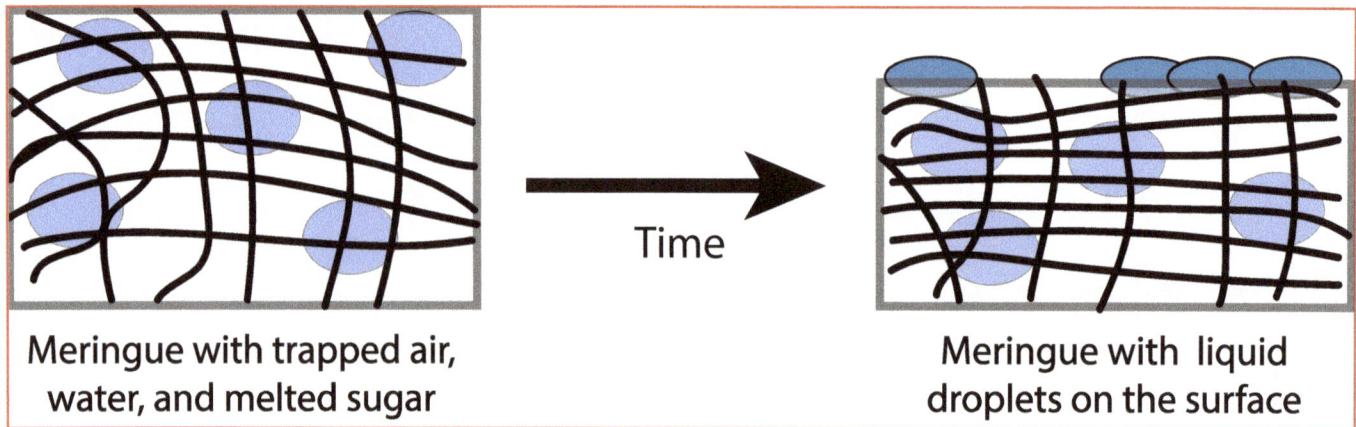

Meringue with trapped air, water, and melted sugar → Time → Meringue with liquid droplets on the surface

Meringue cookie recipes on the internet say you can flavor them with vanilla, almond extract, peppermint, cinnamon, orange juice, or fruit jam before baking. You can even fold in freeze-dried fruit (like raspberries) or chocolate chips, or add food coloring. Look at all these delicious photos!
https://is.gd/gfLEUu.

Your imagination and desire to experiment is the only limit. That's it! Now let's clean up the mess!

MESS #4: EGG-FOAM-TASTIC! 35

Lemon Meringue Pie[11]

MESS 5: GREEN EGGS THAT SPIN!

Background

A popular way to eat eggs is boiled. Let's take a look at a common problem with boiled eggs: weird-colored yolks!

Items Needed

- 4 eggs (in the shell, unbroken)
- 1 large saucepan
- 1 large mixing bowl with water and ice
- 1 large spoon with slots (for getting eggs out of boiling water)
- 1 pencil
- 1 plate
- Masking or painter's tape for labels
- Marker or pen

Procedures

1. Fill the bowl with water and ice.
2. Place all 4 eggs in the saucepan and fill it with water to about 3/4 full. We need it this full since lots of water will be lost while boiling.
3. Carefully place all the eggs in the saucepan.
4. Bring the water to a boil, making sure to watch for boiling to begin.
5. Once boiling starts, set a 10-minute timer. At 10 minutes, use the slotted spoon to remove one egg and place it in the ice water bowl. Once it's cooled, mark this egg "10" with the pencil.

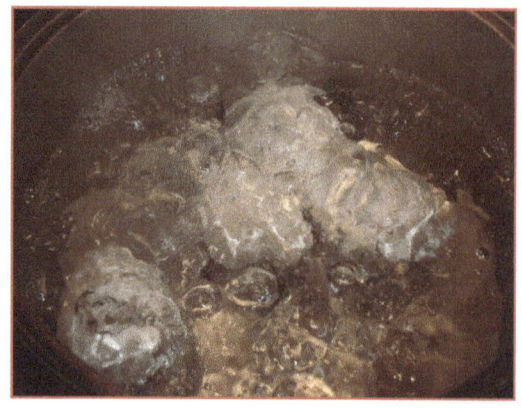

36

6. Immediately set the timer for 5 minutes and continue boiling. After the 5 minutes, remove one more egg and place it in the ice water bath. Mark this egg "15" once it's cooled.

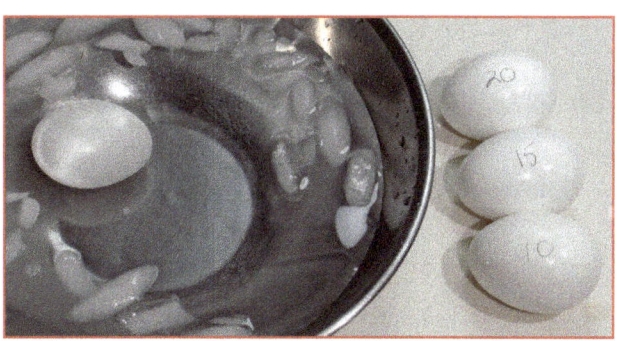

7. Repeat two more times, marking the last two eggs "20" and "25" after they've been removed at 20 and 25 minutes from the boiling water and cooled.
8. Divide the plate with tape, as shown, and mark each plate section with "10," "15," "20," and "25" so you don't get the peeled eggs mixed up.
9. Peel all the eggs and remove the whites, keeping the yolks intact.
10. Observe the yolks. Are they any different? Record your observations.

The Edible Knowledge

Green eggs, anyone? Yep, you should have noticed that the yolks went from a nice yellow at 10 minutes to a light shade of green on the yolk surface that was boiled for 25 minutes. What happened?

Well, there's iron (Fe) in egg yolks and sulfur (S) in the egg whites. Sulfur is what makes boiled eggs smell like they do… like boiled eggs. When you boil the eggs, the iron and the sulfur—if given enough time and heat—combine to form iron sulfide (FeS), which is green!

Different colored eggs: maybe you've only seen white eggs, but different species of chickens lay different colored eggs. Check it out!

https://www.youtube.com/watch?v=2YW3URfUGy0

The green color starts on the yolk's surface, where the whites and yolk are in contact. Up to a point, the longer you boil the eggs, the greener the yolks become. The color doesn't affect the egg's taste. But it sure doesn't look very good.

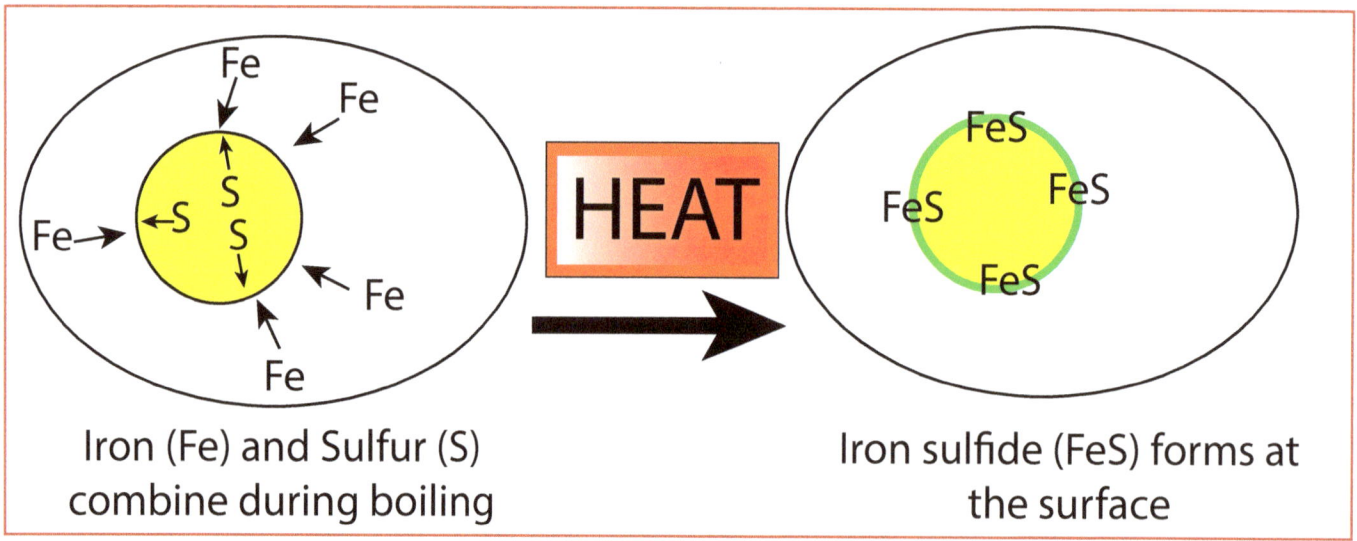

Iron (Fe) and Sulfur (S) combine during boiling

Iron sulfide (FeS) forms at the surface

So what do you do with over-boiled eggs? Go ahead and make them into an egg salad sandwich. It will taste just fine!

That's it! Now let's clean up the mess!

MESS #5: GREEN EGGS THAT SPIN!   39

Hey, wait a minute…

This experiment was called "Green Eggs that Spin!" So where's the spinning?

Here's a fun trick!

If you don't mark your hard-boiled eggs or keep them in a separate container, you can't tell by looking at them which are raw and which are boiled. An easy way to tell the difference is to spin an egg on a flat surface. If it spins nicely, like a top, it's boiled. If not, it's not boiled and is still raw.

Raw vs boiled determined by spinning! https://youtu.be/_3gMxrOwVl

Why do boiled eggs spin? The explanation is a property of physics called inertia. When the egg is raw, the yolk and white aren't really attached to the outside shell. When you spin the shell, the liquid inside resists the spin. The yolk is also denser than the white, making things even crazier as you try to spin it. These facts throw the egg off balance and makes it fall over.

In a boiled egg, the yolk and white are not solid, and for practical purposes, are attached to the shell. Now when the shell is spun, the inside moves as the same speed, and it spins like a top firmly in place in the egg white's middle, allowing the egg to spin nicely.

Now you know!

That's really it!

MESS 6: HUGE MARSHMALLOWS!

Background

Marshmallows are cool and really interesting to make. Unfortunately, they're too complicated for this book, but we are going to play with them. Doing so will let us learn about their structure and a little about something called "leavening."

Items Needed

- 4 regular-sized fresh marshmallows (old ones tend to either dry out or stick together)
- Microwave
- 1 microwave-safe dinner plate

If you want to make s'mores, don't forget the graham crackers and chocolate bars. I'll bet you didn't know you can make s'mores in the microwave!

Procedures

1. Take a close look at a marshmallow. Cut it in half with a sharp knife and examine the interior. Do you see any little air pockets?

2. Place a whole marshmallow on the plate and microwave 5 to 15 seconds, observing closely. All microwaves are not the same, some have much more power, and you can burn a marshmallow this way. Spoiler alert: stop when the expansion starts to slow.

> *S'mores:* If you want to make Smores, you can microwave the marsmallow directly on a graham cracker instead of a plate!

3. What happens during the heating cycle? And what happens when the microwave shuts off? Record your observations:

The Edible Knowledge

Expansion, for sure! Marshmallows are made out of sugar and gelatin—that's pretty much it. Gelatin is a protein, and it allows marshmallows to stretch out and still remain intact. And if you float marshmallows in a cup of hot chocolate, the gelatin helps keep the sugar from dissolving right away.

So why did they inflate so much in the microwave? It's because marshmallows are actually foams, with a stretchy sugar-and-gelatin mix surrounded by small air pockets. Air, which is made up of gases, expands as it heats up. The sugar and gelatin also become hot, which allows them to be stretched easily by the expanding air.

Between the air expanding and the marshmallow itself being flexible, marshmallows grow enormous in the microwave!

You also saw that as soon as the microwave is turned off, the marshmallow immediately starts to shrink. Gas expands and contracts with temperature. When the microwave's heat disappeared, the air contracted, and the marshmallow's weight helped the contraction.

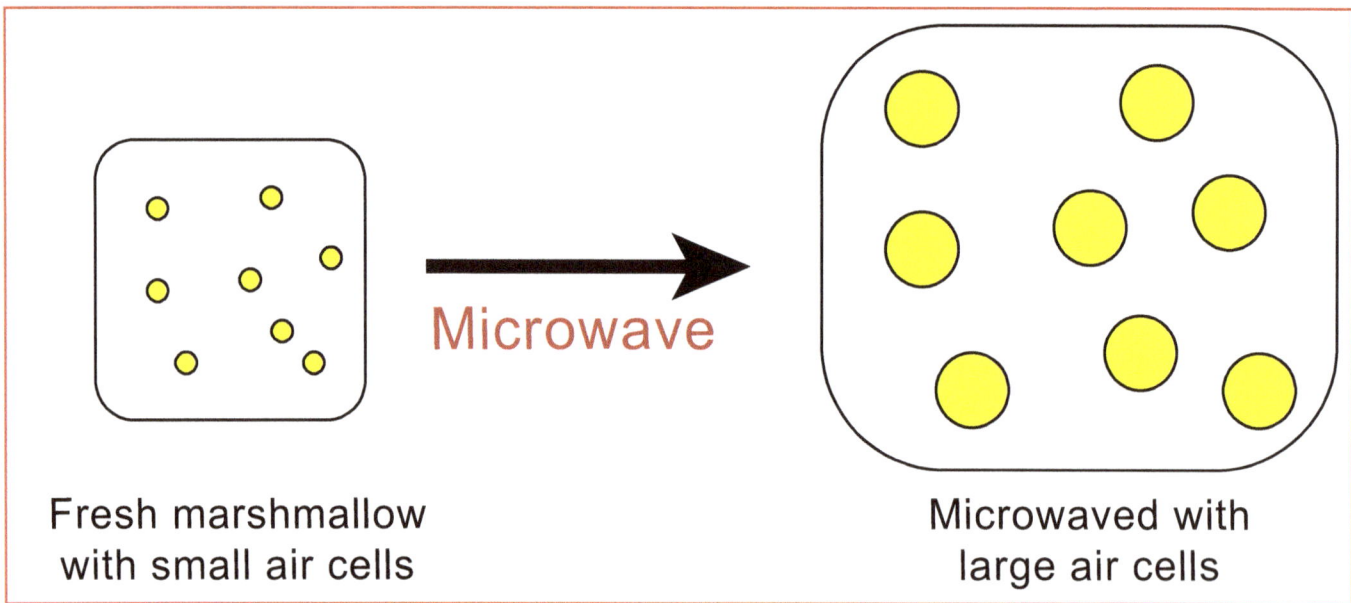

Fresh marshmallow with small air cells → Microwaved with large air cells

Leavening is what makes foods rise when you bake or cook, and is very important in food science. The ingredients that form "leaven" are generally called "leavening agents," and they usually give off a gas. In the marshmallow's case, the gas is already there and trapped. It just expands as it gets hot.

That's it! Now let's clean up the mess!

MESS #6: HUGE MARSHMALLOWS!   43

Wait! What about the s'mores?!

Take a half of a graham cracker, microwave a marshmallow on it, then put a piece of chocolate on the hot marshmallow. Top it with another graham cracker and squish. Yum!

That's it...again!

MESS 7
CAKE FOR DINNER! (OK, MAYBE NOT...)

Background

Cake always means "Celebrate!"

If you have watched someone make a cake, you may have noticed that before a cake pan goes into the oven, the cake batter doesn't fill the pan completely. But when the pan comes out of the oven, the cake batter has expanded and can even be higher than the pan.

How do cakes rise? It's a piece of cake!

In this experiment, let's look at two ways they rise by making two cakes.

No Baking Powder | Baking Powder Added

Items Needed (For Parts 1 and 2)
- See the ingredient lists in each part for specific foods needed
- 1 or 2 9-inch cake pans (round or square)
- 1 electric mixer, handheld or countertop, with beaters
- Measuring spoons
- Measuring cups
- 1 fork or whisk
- 1 small bowl for beating eggs
- 1 spatula
- 2 mixing bowls
- Oven preheated to 350°F
- 1 ruler

MESS #7: CAKE FOR DINNER! (OK, MAYBE NOT…)

PART 1: CREAMED, NO BAKING POWDER

Use the ingredients listed below in a simple cake recipe to make cakes for Part 1 and Part 2.

Ingredients for Part 1

- 2 large eggs (in the shell, unbroken)
- 1/2 cup butter (1 stick), slightly softened, plus more for greasing the pan
- 1 teaspoon vanilla
- 1/8 teaspoon salt
- 1 cup flour
- 1/2 cup milk
- 1/2 cup sugar
- 1/4 cup butter (1/2 a stick), slightly softened

> *Be careful!* Don't simply repeat the first recipe without reading Part 2. The recipes are slightly different!

> *Greasing* pans is often a necessary task in the kitchen. Some cooks spread the butter around with their fingers. Those who don't want to get their hands buttery use a napkin or clean kitchen towel to swipe off a blob of butter and spread it around the pan. You can also pick up one end of the butter with a napkin or clean towel and use the other end to spread it around. It's like coloring with a fat marker.

Procedures

1. Take 1 cake pan and grease all the sides with butter.

2. Measure the flour and salt into a mixing bowl. Blend them together with a fork or whisk.
3. Add the butter and sugar to the **second** mixing bowl. Blend those with the mixer for just 30 seconds, or until moistened.

46 INTRODUCTION TO FOOD SCIENCE *FOR KIDS!*

4. Continue blending for 4 minutes on medium to high speed. This step is sometimes referred to as "creaming." The mix's texture and color changes to a smooth and creamy white or pale yellow.

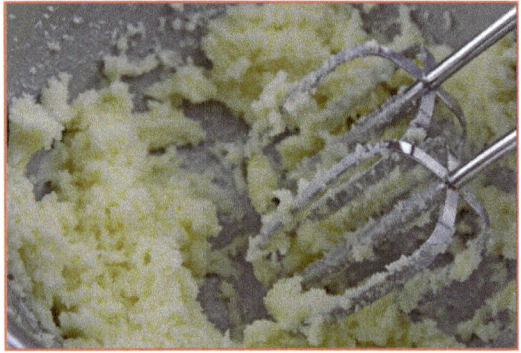

5. Break the 2 eggs into the small bowl and beat until mixed together.

6. Add the milk, vanilla, 2 tablespoons of the beaten eggs, and about half of the flour mixture to the creamed butter and sugar. Mix for about 30 seconds on low.

7. Add the remaining flour and mix for another 30 seconds, just until smooth. If you over-mix the batter, your cake

MESS #7: CAKE FOR DINNER! (OK, MAYBE NOT…) 47

will be tough. (We'll talk about that in the "Gluten in Our Food" experiment.)

8. Pour the batter into the greased cake pan and smooth.

9. Bake in the preheated oven, on the center rack, for about 30 minutes, or until a toothpick inserted into the cake's center comes out clean. Be careful with the hot oven!
10. Remove the pan from the oven and let the cake and pan cool. If you have another 9-inch pan, you can skip right to Part 2. If not, let the cake cool and remove it from the pan.
11. Once the cake is out, wash the pan and all your equipment. Go to Part 2

PART 2: CREAMED, WITH BAKING POWDER

Start with all clean and dry utensils. The object here is to start with the same equipment, in the same condition, all three times. Otherwise, the different conditions will make a comparison impossible.

Cake removal: Letting the cake cool to room temperature makes it much easier to remove from the pan, but if it's stuck, try these suggestions! https://www.wikihow.com/Fix-a-Baked-Cake-Stuck-to-the-Pan

Ingredients for Part 2
- Small bowl with beaten eggs from Part I
- 1/2 cup butter (1 stick), slightly softened, plus more for greasing the pan
- 1 teaspoon vanilla
- 1/8 teaspoon salt
- 1 cup flour
- 1/2 cup milk
- 1/2 cup sugar
- 1/4 cup butter (1/2 a stick), slightly softened
- **1 teaspoon baking powder**

48 INTRODUCTION TO FOOD SCIENCE *FOR KIDS!*

Procedures Part 2

1. Take 1 cake pan and grease all the sides with butter.
2. Measure the flour, salt, and **baking powder** into a mixing bowl. Blend them together with a fork or whisk.
3. Add the butter and sugar to the second mixing bowl. Blend those with the mixer for just 30 seconds, or until moistened.
4. Continue blending for 4 minutes on medium to high speed. This step is sometimes referred to as "creaming." The mix's texture and color changes to a smooth and creamy white or pale yellow.
5. Add the milk, vanilla, 2 tablespoons of beaten egg, and about half of the flour mixture to the creamed butter and sugar. Mix for about 30 seconds on low.
6. Add the remaining flour and mix for another 30 seconds, just until smooth. If you over-mix the batter, your cake will be tough. (We'll talk about that in the "Gluten in Our Food" experiment.)
7. Pour the batter into the greased cake pan.
8. Bake in the preheated oven, on the center rack, for about 30 minutes, or until a toothpick inserted into the cake's center comes out clean. Be careful with the hot oven!
9. Remove the pan from the oven and let the cake and pan cool. Remove the cake from the pan and place it on a plate.
10. Compare the two finished cakes on separate plates, side-by-side.

Write down what you've seen so far:

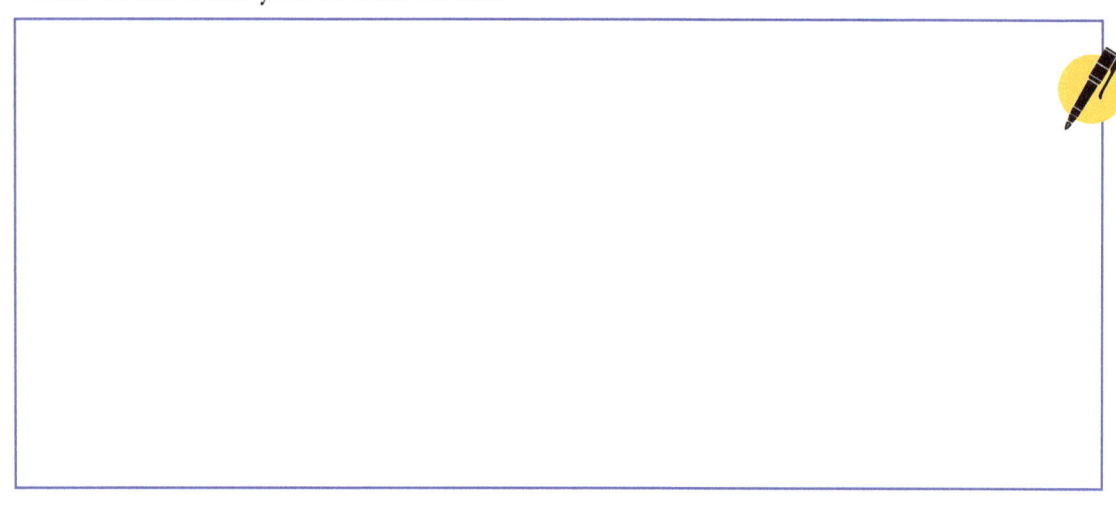

MESS #7: CAKE FOR DINNER! (OK, MAYBE NOT...) 49

OK, now we have two cakes to evaluate! Once both cakes are at about room temperature, cut each in half across the middle, and measure the height at the highest point for each.

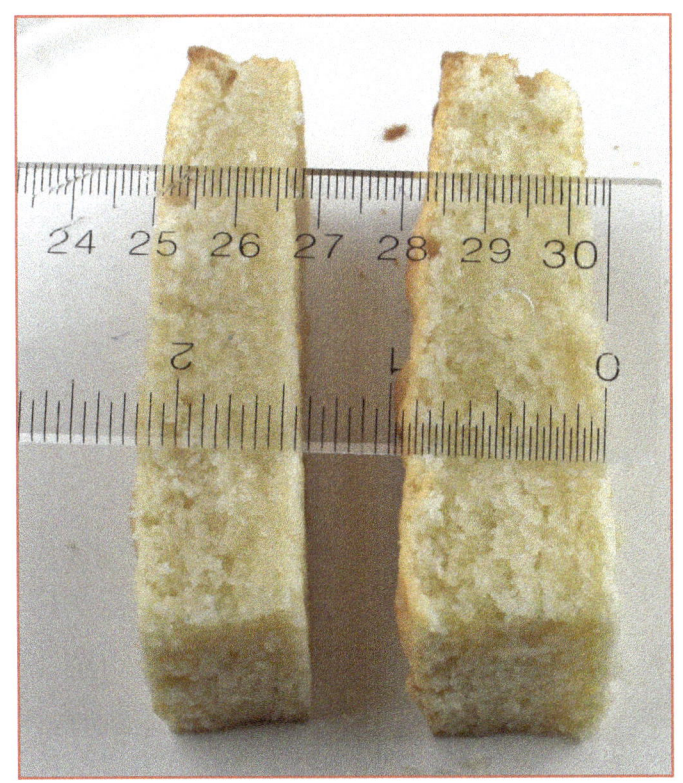

Record your results:

	Height	Texture
Creamed		
Creamed + Baking Powder		

The Edible Knowledge

You likely saw a pretty big difference in the finished height of these cakes. When fat and sugar are beaten together, air gets trapped in the mixture in a special way. Why?

Sugar crystals have sharp edges. Air gets trapped on one or more of these edges. This locked-in air expands when heated in the oven, lifting—or leavening—the cake.

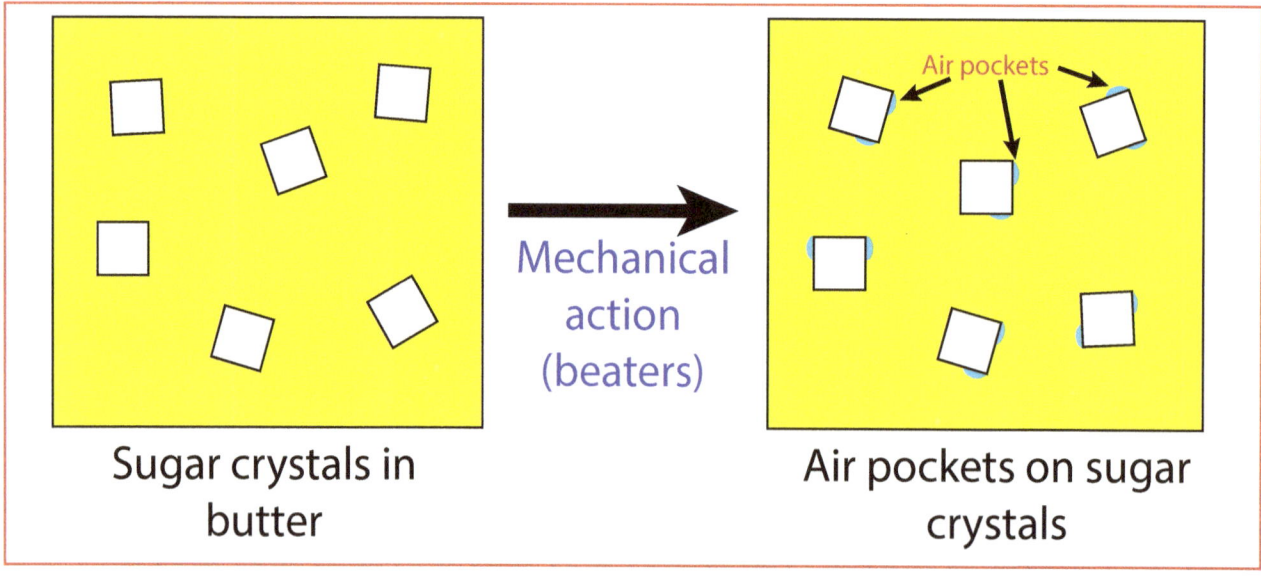

While great cakes can be made this way, it is tricky and requires skill and experience. If you don't mix it fast enough, or long enough, or if it is too warm or too cold, you can end up with a flat cake. Did you?

To help more people make good cakes at home—with baking recipes that are more predictable—baking powder is added. Baking powder contains ingredients that produce gas when heated. Because it makes its own gas, we don't have to worry about beating it in during the creaming step.

Now, make some frosting and eat some cake!

That's it! Now let's clean up the mess!

Simple cake glaze recipe:
In a medium bowl, stir 3 tablespoons milk into 1 1/4 cups confectioners' sugar. Add 1/2 teaspoon vanilla. Drizzle on your favorite dessert.

MESS 8: GLUEY, GOOEY GLUTEN... IN OUR FOOD?

Background

You may have seen products in the store that have the words "gluten-free" on them but don't know what they mean. Or maybe you know more about this issue because someone you know has problems when they eat gluten.

So what is gluten, and why is eating it a problem for some people? It's a problem because some people have an illness called "celiac disease." People with this illness have a lot of problems with their digestive system (stomach, small intestines, and large intestines are the usual organs involved) if they eat something with gluten in it, including days of pain and discomfort.

Others have what's called a "sensitivity," which means while they don't have immediate physical pain, eating gluten can cause other symptoms, such as a rash, joint pain, or other things.

So if you're having a party and feeding people, you need to remember that gluten can be a problem for some people. For the rest of us, it's something we eat every day, with no issues.

Where in the world is the gluten? Gluten comes from wheat, barley, and rye, and those are easy to avoid if it bothers you. Raw fruits and vegetables are naturally gluten-free.

But a lot of the time, it hides in prepared foods, like breaded foods (such as fish sticks), canned baked beans, canned soups, cereals, cold cuts, commercially prepared chocolate milk, Communion wafers, energy bars, French fries, fruit fillings and puddings, gravy, hot dogs, ice cream, ketchup, mayonnaise, meatballs, meatloaf, root beer, salad dressings, sausage, trail mix, and veggie burgers.

Fortunately, since it is a problem and since gluten can hide under different names, bold print at the bottom of a nutrition label will say "Gluten-Free" if the product meets requirements. If gluten bothers you and a food label doesn't say "Gluten-Free," leave it alone.

Wow!

MESS #8: GLUEY, GOOEY GLUTEN… IN OUR FOOD? 53

Note: Some adults have purchased an electric bread-making machine, but don't use it. This experiment has to be done by hand.

That's good because gluten makes for great bread! Let's look at what gluten is and the function it plays. It's some gluey, gooey fun… We're going to make some bread!

Items Needed (For Parts 1 and 2)
- 2 loaf pans (should be the same size for best results)
- Oven preheated to 375°F
- 1 large mixing bowl
- A spoon and a whisk for mixing

Here's the recipe, <u>which is the same</u> for both Part 1 and Part 2:
- 1 cup water
- 1 tablespoon instant or active dry yeast
- 1 1/2 teaspoons salt
- 2 tablespoons sugar
- 2 tablespoons vegetable shortening (like Crisco®, solid at room temperature)
- 2 1/2 cups all-purpose flour

PART 1 – VERY LITTLE KNEADING

Procedures
1. Add the water, salt, sugar, shortening, yeast, and 1 cup of flour to the mixing bowl.
2. Use the whisk to blend the ingredients until the shortening has been mostly broken up.

3. Add the rest of the flour (another 1 1/2 cups) and stir with the spoon until just mixed.

4. Add to a greased (butter or solid shortening) loaf pan and set it aside to rise (through yeast action) while you work on the next loaf. Both loaves will go into the oven together.

PART 2 – LOTS OF KNEADING

1. Follow Steps 1 and 2 in Part 1.
2. Add the rest of the flour (another 1 1/2 cups) and continue mixing with the spoon until the dough starts to "follow" the spoon.

3. Remove the dough and place it on a floured surface, like a table or countertop.
4. Add a small amount of flour to the dough's surface, flour your hands, and begin *kneading*. The dough will stick to your hands and the surface, so don't worry, that's normal!
5. Continue kneading and adding small amounts of flour for at least 10 minutes. At some point during this time, the dough becomes stretchy and springy, and it will cling to itself rather than the table or your hands.
6. Form the dough into a loaf and place in the second greased pan.

Kneading dough: If you've never kneaded dough before, here's a video that shows you how: https://www.youtube.com/watch?v=5ZqJyalYqKU

7. Let the 2 pans rise for 60 minutes.

8. Place both loaves in the preheated oven and bake for approximately 40 minutes, or until the loaf that was kneaded sounds hollow when tapped. Remove them from the oven and let cool.

MESS #8: GLUEY, GOOEY GLUTEN… IN OUR FOOD? 57

9. Slice a piece off each. Observe the appearance, texture, and taste.

Record your observations regarding the differences between the loaves:

The Edible Knowledge

Gluten is formed from two kinds of protein found in wheat, called "glutenins" and "gliadins." When they combine—which is what happens when they run into each other—they form gluten, which is a very stretchy material. The purpose of kneading is to help the glutenins and gliadins collide so they can form gluten.

If not enough gluten is formed, a loaf of bread might rise, but then it will collapse in on itself. Or it might not rise much at all.

The reason is that the stretchy gluten traps gases released from the yeast action. We already know gases expand as they warm up in the oven. In this case, they further cause the loaf to rise. Finally, the gluten firms up once the loaf comes close to being done, stabilizing the bread's volume. The finished bread is more crumbly when it doesn't have enough gluten, but it still tastes good!

So Part 1 resulted in a really weird "loaf." Part 2's loaf should seem pretty familiar, although still quite different from store-bought bread.

Gluten development is also why some things, like biscuits or cake, shouldn't be over-mixed. If you do, you'll end up with more of a bread-like texture, which doesn't make for good birthday cake.

That's it! Now let's clean up the mess!

MESS 9: BANG PUFFING!

Background

Hopefully you've had the pleasure of eating popcorn. It's one of my favorite snacks! Have you ever wondered how it works? How do we get a light and airy thing we can eat from a hard kernel of corn that's not-so-much-fun to eat? Let's find out.

And you get to eat some breakfast cereal!

Items Needed

- 1 packet of microwave popcorn
- 1 box Post® Honeycomb®[2] cereal
- 1 box Cap'n Crunch's Crunch Berries®[3] cereal
- 1 dinner plate
- 1 sharp knife or utility knife

Procedures

1. Pop the microwave popcorn according to the instructions.
2. Take a few pieces each of the popcorn, Honeycomb®, and the "berry" part of Crunch Berries® and place them on the plate.
3. Using a sharp knife, cut several pieces of each in half.

60

MESS #9: BANG PUFFING! 61

4. Now look at the insides of the pieces. Describe what you see:

5. Eat a few. What's the texture like of each? Record your observations:

The Edible Knowledge

Popcorn, Honeycomb®, and Crunch Berries® all come from the same basic food element, but they are created by different processes.

Remember starch? All these products are primarily made from starch. In the case of popcorn, it's cornstarch—of course! Starches react with heat. When they reach a specific temperature, the starch becomes rubbery and pliable. It's a pretty high temperature, so you have to heat it for a while. Before that, the starch is brittle and called "glassy." The temperature at which starch changes from glassy to rubbery is called the "glass transition temperature."

OK, so how do we get from that to a puffed piece of food that's light and airy and fun to eat?

When starch is contained and heated until it becomes rubbery, and then the pressure is suddenly released, the pliable starch expands rapidly. At the same time, the starch quickly cools below the glass transition temperature and firms up.

Another way of achieving this same explosive-and-instant-expansion is to create pressure inside a cooker, then release it with a valve. The cooker acts like the shell, which "bursts" when the valve is suddenly opened. Just as for popcorn, it makes a loud sound, but much louder. Let's look at an example: Honeycomb®.

A dough is made and formed into the shape we know as Honeycomb® cereal by pushing the dough through a press like a Play-Doh®[10] press. These pieces are small, maybe 3/8" in diameter, and they are all the same size when they are added to the cooker.

They are cooked under pressure, then the valve is suddenly released. BANG! The pieces shoot out! This process is actually referred to as "gun puffing" because it sounds like guns, or canons, going off. I've worked on Honeycomb® and other gun-puffed cereals, and the pressure release noise is really loud!

> **Popcorn** kernels have a hard shell that forms a container around the starch being heated. Water in the kernel turns to steam and builds up a lot of pressure. Eventually, the kernel "shell" will burst, allowing for instant expansion. When the shell bursts, it makes a loud popping sound (popcorn, get it?) when the pressure is released.

A cool video showing popcorn popping in slow motion:
https://www.youtube.com/watch?v=FSZd33awqQk

One interesting thing about gun puffing is that the pieces burst out irregular in size. Look at the picture of Post® Honeycomb®. Believe it or not, they all were the same size when they were loaded into the gun, where they sat in a pile.

Conditions in different areas of the gun vary, though. The pieces in the center of the pile get hot slower than those on the outside. That's one reason why they expand differently when they're shot out!

The last cereal, Cap'n Crunch's Crunch Berries®, are made through a process called "extrusion." In this manufacturing process, dough is first cooked. Then it's pressed against a small opening under tremendous pressure and temperature. When it finally squeezes its way out, it immediately expands.

Gun puffing video: https://www.youtube.com/watch?v=7OoiACNV_tU

A video showing extrusion
https://www.youtube.com/watch?v=3LJto1D0iPs

Pretty cool, right? Now open a different box of cereal from your kitchen. Look at a few pieces and try to decide which process was used to create it. If you want to know for sure, the internet probably has the answer.

Now eat some cereal! Just don't spoil your appetite for dinner…

That's it! Now let's clean up the mess!

MESS 10: FERMENTATION IS EVERYWHERE!

Background

Wait a minute… I'm only six! I can't even be left in a car by myself, but I can eat fermented stuff? Isn't that how they make beer?

Yes and yes! But fermentation isn't only for beer. The process is needed for many food products we eat. Let's look at a few.

Whoever is supervising this experiment should like it. It's a little simpler, with mostly tasting and discussion.

Fermentation is what happens when bacteria, yeasts, and molds eat some of the sugar in our food. The process results in byproducts (yes, their "waste" as they eat!) that bring about a welcome change to food or a drink: wine and beer, and sour foods like pickled cucumbers, kimchi, and yogurt. Almost every continent on Earth features some type of local fermented food.

MESS #10: FERMENTATION IS EVERYWHERE! 67

Items Needed

- Cheddar cheese
- Sourdough bread
- 1 piece of chocolate
- 2–3 slices of pepperoni
- 4 small plates
- 1 sharp knife

Procedures

1. Slice off 2–4 pieces of cheddar cheese, sourdough bread, and pepperoni, and place them on a plate. Put a couple of pieces of chocolate on another plate.
2. Look carefully at each and taste them, trying to distinguish anything that might be common among them. Note: Taste the pepperoni last, as it has a strong flavor that will stay in your mouth. If you eat it first, you won't be able to evaluate the others accurately.
3. Fill out the below table with your observations.

	Cheddar Cheese	Soudough Bread	Chocolate	Pepperoni
Flavor				
Texture				
Appearance				

The Edible Knowledge

These four foods get their characteristic flavors from fermentation. Some are also preserved that way. All bacteria produce byproducts as they live—eating, splitting, and what would be their equivalent of pooping and peeing.

Just like us!

These byproducts are sometimes useful, and sometimes full of flavor that humans like. Or, at least, we have learned to like them! Let's look at each one.

Cheddar Cheese

Cheese[4]

Cheese is made by using a "starter culture" that contains types of bacteria that will end up in the desired result. The starter culture is mixed with milk and left to live for a while at conditions that work well for the bacteria, usually a warm temperature. Among other things, these bacteria produce acid, which has the result of making the milk set up, like a gel.

I've actually participated in cheese production on a large scale, and it's fascinating. A huge vat of milk turns to a gel in a couple of hours, without changing its appearance really at all. When you touch it, though, you know something happened. It's no longer liquid, but firm and jiggly, like touching a gelatin dessert, but it's warm.

An interesting video showing cheese production:
https://www.youtube.com/watch?v=y9wLhRrj5Ug

> **What happens** to all that whey? Cheesemakers can use some of it to make ricotta cheese, and they can sell it as animal food. It's also used in a lot of other food products.

The gel is holding a lot of liquid, then it's cut into pieces by running what looks like combs made with wire through it. What's left are two distinct products: the curds, the more solid part, and the whey, which is liquid. The whey is drained off and the curds are squeezed to remove even more whey.

The resulting pressed "loaves" of curds become cheese. If you look closely at some thin-sliced cheddar cheese, especially if you hold it up to a light, sometimes you can see the individual outlines of curds. The cheese curds are also salted, which slows down and limits the bacteria's continued activity, but doesn't stop it.

Aging is an important step, during which the bacteria continue to function and change the flavor. Some like their cheddar cheese sharp (older) and some like it mild (younger).

Pretty cool!

Sourdough Bread

Sourdough bread[5]

You may not have tried sourdough bread before. It's one of my favorites because I love the flavor and texture.

Like cheddar cheese, a starter culture is used, sometimes referred to as the "mother dough." A small amount of dough that contains the active bacteria is mixed with fresh flour and other ingredients and allowed to react for a while. Acid is produced by the bacteria, which is what gives the bread its characteristic sour flavor. Other byproducts and the bacteria's action affect the bread's finished texture.

Baking kills all the bacteria, so bakeries continually save portions of the dough, the "mother dough," to start the next batch. Bakers guard their mothers carefully, and some have been in existence for generations, always being preserved and passed down.

Isn't that amazing?

You can buy lots of sourdough breads with various flavors and textures. Apart from the ingredients and varying methods of mixing and baking, the particular starter cultures are what make the difference.

Different colored eggs: maybe you've only seen white eggs, but different species of chickens lay different colored eggs. Check it out!

https://www.youtube.com/watch?v=2YW3URfUGy0

Acid and sourness: Sour candy has crystalline stuff on the outside that makes them sour. This is usually citric acid, which comes from citrus fruits, among other sources.

Chocolate

Chocolate[6]

Did you know that chocolate is a fermented product, or at least made from fermented products? Cocoa beans are what give chocolate its flavor, but in their natural form, they don't taste at all like what has become a treat enjoyed all over the world.

MESS #10: FERMENTATION IS EVERYWHERE! 71

The beans grow in pods that also contain a lot of wet, pulpy material. When harvested, the beans are scooped out with the pulp and left in piles in a location outside. With lots of protein, carbohydrates, and water, these piles are perfect breeding grounds for the bacteria that cause composting and fermentation, which is exactly what happens.

The length of time and conditions during fermentation change the cocoa bean flavor to what we expect. The beans are eventually washed and dried and sent to "chocolatiers," or chocolate manufacturers. The process from that point is still a long one, and won't be covered in this workbook.

Interestingly, "chocolate" is definitely not the same around the world. I lived for a while in Mexico. The chocolate there wasn't something I enjoyed, as the texture and flavor were quite different. You might hear about Swiss chocolate, which can be very different than a HERSHEY'S Milk Chocolate bar!

All chocolate is good, but we like what we become accustomed to. Don't be afraid to try something new, though. You might find you really like it, and the same is true for more than chocolate!

Making chocolate: A great video showing chocolate production from the field all the way to the bar. https://www.youtube.com/watch?v=V-4FsJ6-bzc

Pepperoni

Pepperoni pizza[7]

Lastly, we have pepperoni. Have you noticed how most pepperoni is sold? You can find sliced pepperoni in bags hanging in store aisles. Pepperoni is also sold sausage-like, in long tubes that may not need to be refrigerated before they're opened (they should be refrigerated after opening). Some pepperoni packages do require refrigeration, so follow the instructions on the package!

How can a meat hang in an aisle, at room temperature, without rotting? Fermentation does it again! The meats and seasonings in pepperoni are mixed together with another starter culture of bacteria. The bacteria produce acid and other byproducts that provide some of the characteristic flavor. The acid eventually changes the environment enough that pathogenic bacteria (bacteria that can kill you), or other bacteria or fungi that can spoil the meat, can't survive or function. That's why it can hang out in a store aisle.

A video showing pepperoni being made: https://www.youtube.com/watch?v=1MxYWwouFPM

So there—you eat fermented stuff all the time and didn't know it. Now enjoy some more of your favorite. I'll bet you're going to go for the chocolate!

That's it! Now let's clean up the mess!

EATING BROWN APPLES? YUCK!

Background

Brown apples are fine…if it means they're covered in caramel! Otherwise, something has gone wrong. A very common problem is that when you cut up apples, they quickly turn brown on the surface.

But have you ever purchased pre-cut apples from a fast-food restaurant? Those stay nice and white!

Let's see how that's done…and what's better, you can also do this at home.

Items Needed

1. 1 or 2 medium-sized apples (it's fun to try this experiment with two different kinds of apples), **room temperature—NOT cold**
2. Water
3. Calcium ascorbate capsules (these can be purchased at grocery stores, or online)   1 large dinner plate
4. 1 medium-sized bowl
5. Masking or painter's tape for labels
6. Marker or pen

MESS #11: EATING BROWN APPLES? YUCK!

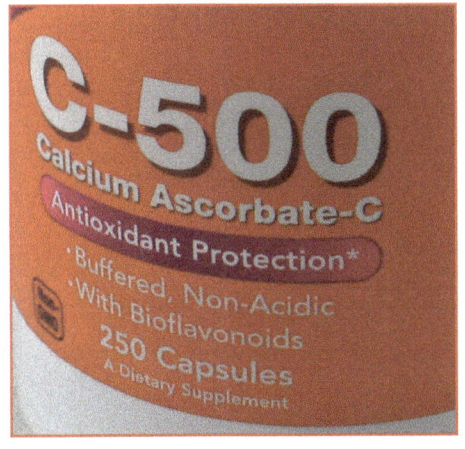

Procedures

The browning reaction will begin right away, so you want to treat them immediately. Please read through the procedures and prepare everything that's needed <u>before cutting the apples.</u>

1. Divide the plate down the middle with painter's tape. Label one side "Plain" and the other "Treated."
2. Prepare the calcium ascorbate dip: open 12 capsules and dump the capsule contents into 2 cups water. Stir well, let it for 5 minutes, stir again. Any material still floating can be skimmed off and thrown away.

3. Cut the apple into about 3/4-inch pieces, with the core removed.

4. Immediately place half the pieces in the ascorbate dip. Stir gently and let them sit in the dip for 5 minutes.

5. While waiting for the dipped apples, place the non-treated apples on the plate's side labeled "Plain."
6. Remove the apples from the dip after 5 minutes and put them on the plate's side labeled "Treated." <u>Record the time.</u> This will be the starting time for your test.
7. Observe the fruit at 0 minutes, 30 minutes, 2 hours, 4 hours, 6 hours, and 24 hours.

You may want to take a picture as you start to help you remember what they looked like at "time zero." You may even do this at other points throughout the experiment. Record your observations.

Time	Plain	Treated
0 Minutes		
30 Minutes		
2 Hours		
4 Hours		
6 Hours		
24 Hours		

The Edible Knowledge

Enzymes are very important proteins in life. Our own bodies make thousands that literally keep us moving all day long! Just as in our bodies, they're present wherever life is, including in apples.

One enzyme, called "polyphenoloxidase" (say poly-fe-no-lox'-i-dace) is important to the apple's growth. But it also makes the surface of cut apples turn brown when exposed to oxygen! While brown cut apples are perfectly safe to eat, most people don't think the brown surface looks good.

So while you CAN eat it, and it won't taste much different, you might not want to.

Fortunately, you can defeat enzymes!

<insert superhero music here>
https://www.youtube.com/watch?v=M7ghouZQRe4

Ok, back to business.

One way enzymes function is by bringing molecules together so they can react easily. Enzymes attach themselves to the molecules, kind of like a puzzle. Each type of enzyme has a particular structure, and that surface will only adhere to a molecule that matches. Think of a square peg and a round hole. The square peg won't go into the hole because it doesn't match.

Enzyme puzzles can be defeated by filling the spot where a puzzle piece is supposed to go. If you have a round peg and a round hole, filling the hole will prevent the peg from entering. That's what we did with the calcium ascorbate.

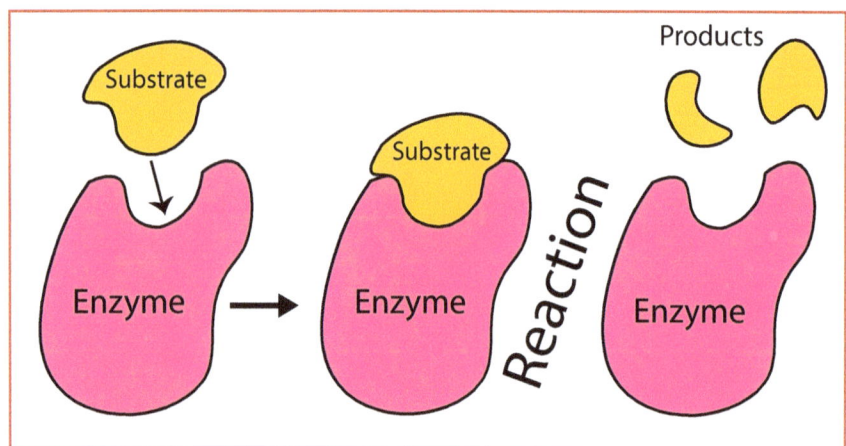

Enzyme Terms: A *substrate* is what the enzyme uses and changes. If an *inhibitor* is present, it can plug the hole in the enzyme, preventing the substrate from entering, then the change, or *reaction*, doesn't happen.

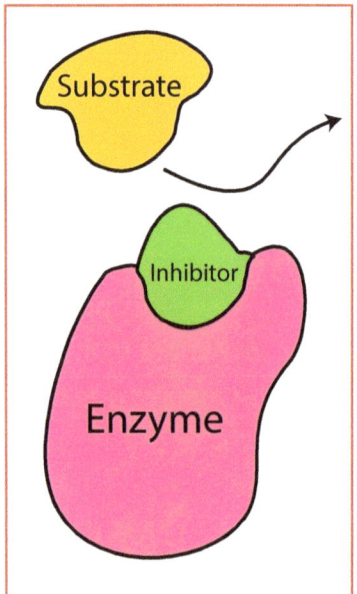

And what's better, calcium ascorbate is related to ascorbic acid, better known as vitamin C! Food science lets you figure out how to make food look and taste better, and be better for you!

MESS #11: EATING BROWN APPLES? YUCK!   79

If you got to try the experiment with different kinds of apples, you may have noticed they browned differently. Each type of apple's composition (what it's made of) is slightly different. Some brown more than others. Pretty cool!

That's it! Now let's clean up the mess!

MESS 12: WATER IS ACTIVE!

No, no, no. Not active water! Water **is** active. Read on!

Background

A very important concept in food science is something called "water activity." It's a measure of how available water is in a food. Water activity and water content are two different measurements.

Here you'll see one essential aspect of water activity. Let's get started!

Items Needed

- Corn flakes from a new box
- Raisins (fresh, unopened container)
- 3 zip-closure–type plastic bags, freezer variety (thicker plastic), quart- or gallon-sized
- 1 cup measuring cup

Procedures

1. Add 2 cups of corn flakes to 2 of the bags, leaving 1 empty.
2. Add 30 raisins to 1 bag with corn flakes, and about 1/4 cup raisins to the empty bag. You should now have 1 bag of corn flakes, 1 bag with corn flakes and raisins, and 1 with raisins. Zip them all closed and set them aside at room temperature.

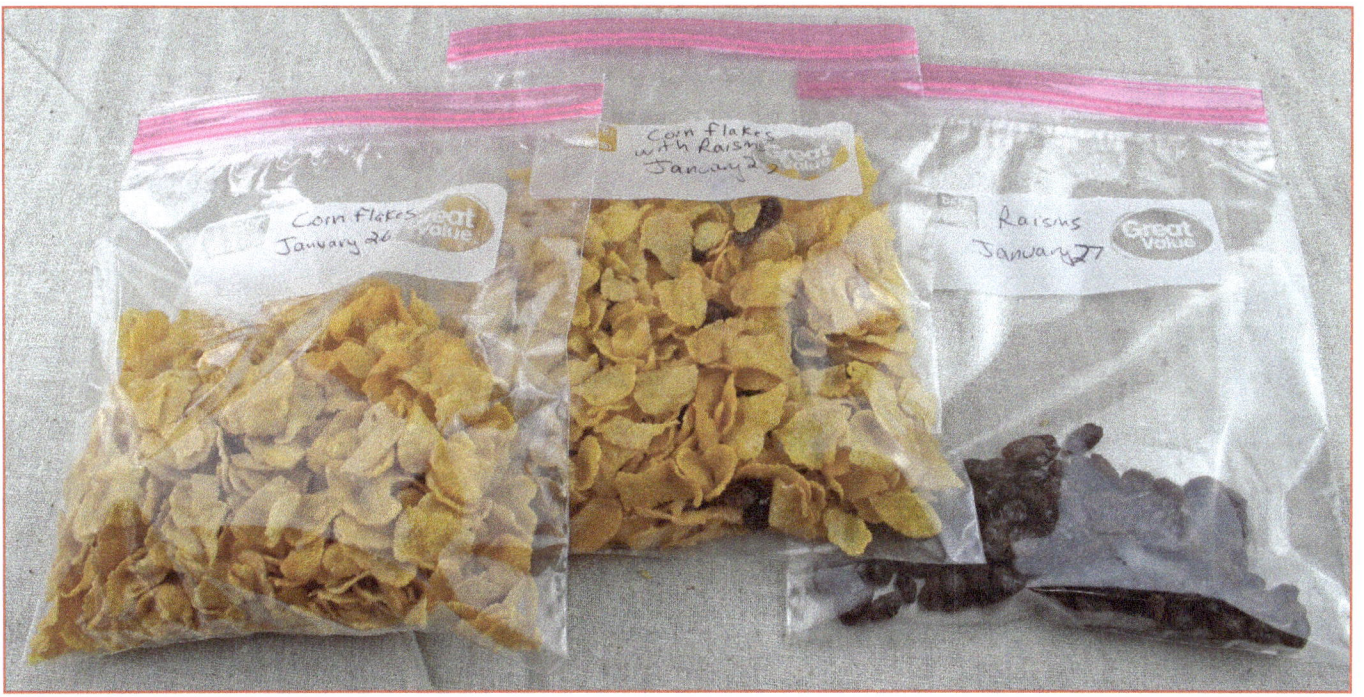

3. Test by tasting some of your extra corn flakes and raisins and fill out the table below.

Fresh (Time Zero)	Texture: soft (1) to hard (5)	Other Observations
Raisins		
Corn Flakes		

That's it for now! Set your bags aside where no one will open them…or eat the contents! Put an X on your calendar 7 days from today and wait until those days pass.

82 INTRODUCTION TO FOOD SCIENCE *FOR KIDS!*

After 7 Days

Take some pieces from each bag, but just a few. Leave the rest in the bags and seal them up again. As before, evaluate the pieces of product and fill out the table below.

Time: 7 Days (1 week)	Texture: soft (1) to hard (5)	Other Observations
Plain corn flakes		
Corn flakes from corn flake & raisin combination bag		
Raisins from corn flake & raisin combination bag		
Plain raisins		

Set the bags aside again. Put an X on your calendar 7 days from today and wait until those days pass.

After Another 7 Days (14 days total)

Repeat the testing on each bag and fill out the table below.

Time: 14 Days (2 weeks)	Texture: soft (1) to hard (5)	Other Observations
Plain corn flakes		
Corn flakes from corn flake & raisin combination bag		
Raisins from corn flake & raisin combination bag		
Plain raisins		

Set the bags aside again. Put an X on your calendar 7 days from today and wait until those days pass.

The Edible Knowledge

What's in a food makes a big difference in how easily it gives up water. The actual amount of water—meaning whether it's wet or dry—can also make a big difference.

Food scientists have to keep this in mind, especially if they need to mix ingredients together, like we did. If they aren't careful, some of their ingredients can become hard, like the raisins in this experiment. Other ingredients can become soft.

Moisture moves from one place to another depending on what "wants" it more. The very dry corn flakes "soaked" up moisture from the raisins, which makes the raisins hard. The corn flakes will also be a little damp, but barely noticeable in this experiment.

84 INTRODUCTION TO FOOD SCIENCE *FOR KIDS!*

The wetness in the raisins cannot be seen, but the wetness and its movement can be illustrated.

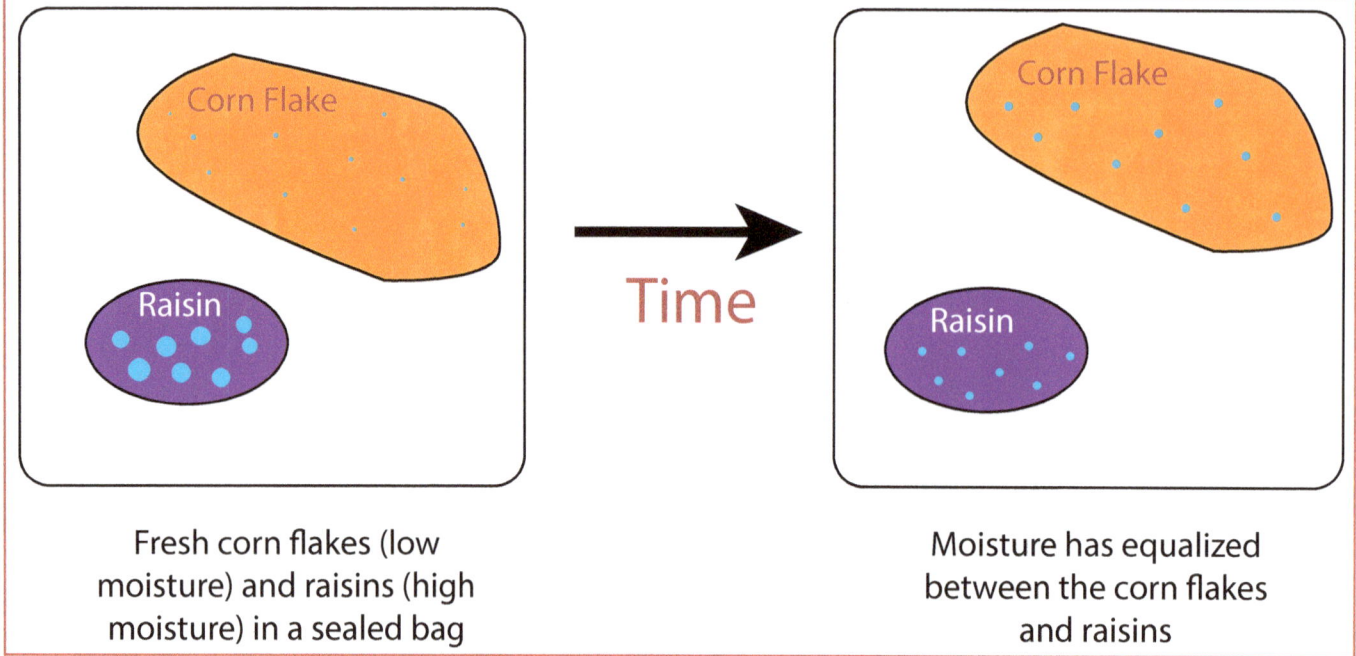

Fresh corn flakes (low moisture) and raisins (high moisture) in a sealed bag

Moisture has equalized between the corn flakes and raisins

To avoid this problem, bran flakes are sprayed lightly with water before the raisins are added in any raisin bran cereal!

There's a lot more to water activity—a LOT more. It's one of the most important concepts in food science! If you're interested in learning more, look at my other Edible Knowledge® workbooks.

***Trail Mix*[8]**: Another example where food scientists make sure water moving between ingredients isn't a problem.

That's it! Now let's clean up the mess!

MESS 13 — HAVE A COW!

Background

We're going to play with beef today. Specifically, we're going to cook beef in two different ways, and explain why we'll end up with two very different finished foods.

This experiment might be good for dinner!

Items Needed

- 3-pound beef roast
- 1 frying pan
- 2 tablespoons olive oil
- 1 oven roasting pan with lid, or aluminum foil to serve as a lid
- Oven preheated to 325°F
- 1 plate

> **Beef** can be expensive. If purchasing beef is not in your family's budget, you can modify this experiment by using chicken breasts. The results won't be as dramatic, but the concepts are the same. Wherever the roast is mentioned in the experiment, simply use chicken breasts instead.

Procedures

1. Preheat the oven to 325°F.
2. Cut a 1-inch–thick slice off the roast, making a "steak." Place the steak—in a bag or on a covered plate—in the refrigerator. It will be prepared later.
3. Place the remaining beef in the oven roasting pan and cover with the lid. If you don't have a lid and are using aluminum foil, try to make it as form-fitting to the pan as possible. You don't want it to dry up.

4. Place the roast in the oven and cook for about 3 hours, but in between, continue on to the next step.
5. About 15 minutes before the baking time's end, heat the frying pan on medium heat, adding the oil to the pan.
6. You can tell when the pan and oil is hot—the oil starts to "shimmer," or look like it's moving. When it's hot, add the steak cut earlier. Sear both sides and continue frying until the center is just slightly red. Remove the meat from the pan and place on a plate.

7. Remove the roast from the oven and then the pan, and put it on the same plate as the steak. Cover the plate with aluminum foil and wait (let *rest*) for about 15 minutes.
8. Cut several pieces off the pan-fried steak and several pieces off the roast. Taste both, and fill out the table.

The "resting" time after removing the steak from the pan and roast from the oven allows for the juices in the meat to move throughout the beef evenly. If you don't let the meat rest, the surface tends to be drier. This resting period is a great chef trick.

Preparation Method	Flavor	Texture	Color
Frying			
Roasting			

The Edible Knowledge

The biggest difference between the two preparation methods is the resulting texture. The steak is tough and chewy, but the oven roast is so tender that it almost falls apart, by itself, into individual fibers. "Collagen" is a connective tissue in meat that holds these muscle fibers together in bundles, and in other ways. Collagen eventually breaks down, but it takes a period of extended heating.

There wasn't enough time for that to happen in the frying pan, so the collagen is mostly still present, making it tough. In the oven roast, the collagen has broken down, resulting in very tender beef.

Oven roasts are my favorite. In fact, I'm getting hungry just writing about it! What is your favorite meal?

You can prepare steak that is tender as well, but it has to come from a muscle without so much connective tissue. One common "cut" of beef for a very tender steak is filet mignon.

Text box: The "resting" time after removing the steak from the pan and roast from the oven allows for the juices in the meat to move throughout the beef evenly. If you don't let the meat rest, the surface tends to be drier. This resting period is a great chef trick.

Now enjoy your beef dinner!

That's it! Now let's clean up the mess!

MESS 14: CREAM THAT'S ICEY AND SMOOTH...AND LUMPY?!

Background

Well, we just had dinner, so now it's time for dessert! We're going to make ice cream in two different ways, learning about crystallization in the process.

Items Needed

- Electric handheld mixer
- 2 3-cup or larger containers with lids
- 1 large mixing bowl
- 1 large spoon for stirring
- 1 spatula or butter knife
- 2 cups heavy whipping cream
- 2 cups half-and-half cream
- 1 cup sugar
- 2 teaspoons vanilla extract
- 1 freezer
- Masking or painter's tape for labels
- Marker or pen

Procedures

1. Label each container with tape, marking one "Stirred" and the other "Not Stirred."
2. In the large bowl, mix the cream, half-and-half, sugar, and vanilla together with the spoon until the sugar is dissolved.

MESS #14: CREAM THAT'S ICEY AND SMOOTH…AND LUMPY?!   91

3. Pour about half of the mixture into each container. Attach the lids and place them in the freezer. The container marked "Not Stirred" shouldn't be touched until the end of the experiment, so be sure to put it in a protected area in the freezer. Then set a timer for 30 minutes.

4. After 30 minutes, remove the container marked "Stirred" and stir it with the electric mixer on medium speed, also using a spoon to make sure all the frozen material is lifted off the sides and bottom.

5. Place the container in the freezer again and set the timer for another 30 minutes.

6. After another 30 minutes, remove and repeat Steps 4 And 5.

7. Continue freezing and mixing until the frozen material begins to get somewhat stiff throughout.

Hint: The product freezes from the container's sides into the middle. It also helps speed up the freezing process if you keep the beaters in the freezer.

8. Remove the beaters, smooth over the top with a spatula or butter knife, and cover it again. This time, let it sit in the freezer undisturbed for at least 3 hours, perhaps longer. Note: The total freezing time will depend on your individual freezer and how it's loaded.
9. Once the time has passed, remove the containers (if they still aren't frozen, put them back in for a while). Use a clean spoon to taste the ice cream in each container and then fill out the table below. You'll be able to feel the ice crystals from both on your tongue. The crystals will all be small (like table sugar), but the tongue is quite perceptive and you'll be able to "feel" differences.

Freezing Method	Texture	Ice crystal size by mouthfeel (small or large)
Stirred		
Not stirred		

The Edible Knowledge

As you know, freezing water forms ice crystals. If the ice cream is left untouched, these become large, even large enough to see them.

There's a lot of water in cream and half-and-half, so it's going to form ice. For ice cream to be nice and creamy, the ice crystals should be very small. That's why when it's made commercially, ice cream is frozen at the same time it's being constantly stirred. The result is ice cream with very small ice crystals. Sometimes they're so small that they aren't detectable on the tongue.

Stirring the ice cream with a handheld mixer isn't the best method, but it's much better than leaving the mixture completely undisturbed. You may still be able to see ice crystals in the stirred container, and almost certainly be able to feel them on your tongue.

Most people don't like the feeling of ice crystals in their mouth when eating ice cream. Perhaps you've experienced this before. It can happen when ice cream melts a bit on the way home from the store and refreezes in your freezer. The product with big crystals is sometimes called "grainy" ice cream.

Mixing also traps air into ice cream as it freezes. A measure of how much air is in it is called "overrun." You need air in ice cream, or it can get very hard.

If you have an ice cream–maker at home, those do a pretty good job of keeping ice crystals small during the freezing process. "Flash freezing," or extremely fast freezing, is the best.

That's it! Now let's clean up the mess!

A fun video showing ice cream production. You'll hear about whipping air into the cream as it freezes! Now you know all about it.
https://www.youtube.com/watch?v=8KaBJlGuw3c

MESS #14: CREAM THAT'S ICEY AND SMOOTH…AND LUMPY?! 95

MESS 15: HEY! PINEAPPLE IS EATING MY JELL-O®?

Background

Remember, enzymes are proteins found in all natural organisms, including ourselves and the things we buy to eat. They're really important for life to continue. But they can also cause food to do unexpected things if you don't take some precautions!

Let's take a look at a fun example.

Items Needed

- 1 large box of gelatin dessert, any flavor (except not sugar-free)
- 1/4 cup diced fresh pineapple
- 1/4 cup diced canned pineapple
- 1 medium saucepan
- 2 large spoons
- 2 containers, preferably clear, that each hold about 2 cups of liquid. These containers will transfer the prepared gelatin to the refrigerator.
- Masking or painter's tape for labels
- Marker or pen
- 1 cutting board
- 1 small knife (not serrated)

96

Procedures

1. Mark one spoon using tape and the marker "Fresh" and the other "Canned."
2. Mark one container "Fresh" and the other "Canned" by using the tape and marker.
3. Cut up the canned pineapple FIRST (before the fresh pineapple). Use the cutting board and knife to make approximate 1/4" x 1/4" pieces. You'll need enough for about 1/4 cup. Add them to the container marked "Canned."
4. Cut the fresh pineapple into 1/4" x 1/4" pieces. Add them to the container marked "Fresh."
5. Prepare the gelatin in the saucepan according to the instructions. Make sure to bring the gelatin and water to a rolling boil and to dissolve all the gelatin granules.

> ***Don't Mix the Pineapple!***
> It's vital not to mix the fresh pineapple with the canned pineapple in any way! Even mixing the juices, on utensils or on the cutting board, will ruin the experiment.

98   INTRODUCTION TO FOOD SCIENCE *FOR KIDS!*

6. Pour about half of the cooked gelatin into each container and stir (using the correctly marked spoon) for about 1 minute each. DO NOT use the same spoon to stir both containers!

7. Place the containers in the refrigerator on a level surface. Let them chill for at least 12 hours.
8. Remove the containers from the refrigerator and write down your observations:

The Edible Knowledge

Fresh pineapple contains enzymes that are natural "proteases," meaning they cut up protein strands into smaller pieces. In fact, they're such good proteases that they're used in meat tenderizers that you can sprinkle on a tough cut of meat. Meat is a protein, and these tenderizers cut right through the protein strands, making it less tough.

In our case, the enzymes were active in the fresh pineapple. Gelatin is a protein, just like tough meat, so the pineapple proteases cut up the strands. Despite our intention of having a nice gelled dessert with fresh pineapple, we end up with pineapple soup instead!

So why didn't the same thing happen with the canned pineapple? Again, the reason is enzymes. These amazing molecules can be destroyed or damaged with heat. Canned pineapple is processed through a high heat step that "denatures," or destroys, the enzyme's ability to function. The enzyme still exists in the canned pineapple, but it's been damaged and can no longer cut up protein strands.

Instead of canned pineapple, you could make your own protease-inactivated pineapple by boiling fresh pineapple. That's actually what I did when developing the experiment, but for sure results, canned is the way to go.

Don't worry, though—you can still eat the gelatin soup. It makes for a nice drink! I used to drink warm gelatin on Boy Scout camping trips. The protein and sugar give you lots of energy!

That's it! Now let's clean up the mess!

-THE END-
OR JUST THE BEGINNING?

We hope you enjoyed this serving of Edible Knowledge! If you did, we invite you to try the rest of the series...this is just the beginning! The other Edible Knowledge books are designed for ages 10 and up and introduce many more food science concepts, but in a lot greater depth than this initial taste. They're still a lot of fun...adults are buying them for themselves! More information can be found on our website, or you can contact us with any questions you may have.

www.beakersandbricks.com

Beakers & Bricks, LLC
PO Box 1014
Asheboro, North Carolina 27204
Phone: 1-888-363-4999
E-mail: info@dalewcox.com

That's it! PICTURE AND ILLUSTRATION ATTRIBUTIONS

Unless otherwise noted, all illustrations and pictures are by Dale W. Cox.

1. Cow Picture; Kim Hansen / CC BY-SA (https://creativecommons.org/licenses/by-sa/3.0); https://commons.wikimedia.org/wiki/File:Cow_(Fleckvieh_breed)_Oeschinensee_Slaunger_2009-07-07.jpg; https://upload.wikimedia.org/wikipedia/commons/8/8c/Cow_%28Fleckvieh_breed%29_Oeschinensee_Slaunger_2009-07-07.jpg. Accessed February 2020.

2. Honeycomb® is a registered trademark of Post Consumer Brands.

3. Cap'n Crunch's Crunch Berries® is a registered trademark of The Quaker Oats Company.

4. Cheese; Guillaume Paumier / CC BY-SA (https://creativecommons.org/licenses/by-sa/3.0); https://commons.wikimedia.org/wiki/File:Cheese_at_the_SF_Chronicle_Wine_competition_Public_tasting_2010-02-20_1.jpg; https://upload.wikimedia.org/wikipedia/commons/8/8f/Cheese_at_the_SF_Chronicle_Wine_competition_Public_tasting_2010-02-20_1.jpg. Accessed February 2020.

5. Sourdough bread; Hillarywebb / CC BY-SA (https://creativecommons.org/licenses/by-sa/4.0); https://commons.wikimedia.org/wiki/File:Sourdough_bread_cooling.png; https://upload.wikimedia.org/wikipedia/commons/d/d7/Sourdough_bread_cooling.png. Accessed February 2020.

6. Chocolate; Mohammed Moosa from UK / CC BY (https://creativecommons.org/licenses/by/2.0); https://commons.wikimedia.org/wiki/File:Cadbury_chocolate_squares.jpg; https://upload.wikimedia.org/wikipedia/commons/b/b1/Cadbury_chocolate_squares.jpg. Accessed February 2020.

7. Pepperoni pizza; User:Apalapala / CC BY-SA (https://creativecommons.org/licenses/by-sa/3.0); https://commons.wikimedia.org/wiki/File:Pepperoni_pizza.jpeg; https://upload.wikimedia.org/wikipedia/commons/1/10/Pepperoni_pizza.jpeg. Accessed February 2020.

8. Trail mix picture; public domain; Evan-Amos / CC0; https://commons.wikimedia.org/wiki/File:Planters-Trail-Mix.jpg; https://upload.wikimedia.org/wikipedia/commons/7/71/Planters-Trail-Mix.jpg. Accessed February 2020.

9. Jell-O® is a registered trademark of the Kraft Heinz Company.

10. Play-Doh® is a registered trademark of the Hasbro Company.

11. Lemon meringue pie; Mum's_lemon_meringue_pie_from_above.jpg: julesderivative work: Mlpearc powwow / CC BY (https://creativecommons.org/licenses/by/2.0); https://commons.wikimedia.org/wiki/File:Mum%27s_lemon_meringue_pie_crop.jpg; https://upload.wikimedia.org/wikipedia/commons/4/4d/Mum%27s_lemon_meringue_pie_crop.jpg. Accessed February 2020.

12. Gravy; Famartin / CC BY-SA (https://creativecommons.org/licenses/by-sa/4.0); https://commons.wikimedia.org/wiki/File:2019-11-28_14_40_22_A_bowl_of_gravy_laid_out_for_Thanksgiving_Dinner_in_the_Parkway_Village_section_of_Ewing_Township,_Mercer_County,_New_Jersey.jpg;